Marine Algae, A Sustainable Food Resource

SCIENCES

Agronomy and Food Science, Field Directors –
Jack Legrand and Gilles Trystram

Food Chains, Subject Head – Thomas Croguennec,
Catherine Renard, Véronique Verrez-Bagnis,
Sylvie Swyngedau Chevallier and Véronique Sante-Lhoutellier

Marine Algae, A Sustainable Food Resource

From Traditional Use to the Food of the Future

Joël Fleurence

iSTE

WILEY

First published 2026 in Great Britain and the United States by ISTE Ltd and John Wiley & Sons, Inc.

ISTE Ltd
27-37 St George's Road
London SW19 4EU
UK

www.iste.co.uk

John Wiley & Sons, Inc.
111 River Street
Hoboken, NJ 07030
USA

www.wiley.com

British Library Cataloguing-in-Publication Data
A CIP record for this book is available from the British Library
ISBN 978-1-78945-229-7

ERC code:
LS8 Environmental Biology, Ecology and Evolution
LS8_13 Marine biology and ecology
LS9 Biotechnology and Biosystems Engineering
LS9_5 Food biotechnology and bioengineering
LS9_6 Marine biotechnology and bioengineering

The manufacturer's authorized representative according to the EU General Product Safety Regulation is Wiley-VCH GmbH, Boschstr. 12, 69469 Weinheim, Germany, e-mail: Product_Safety@wiley.com.

Contents

Preface

Algae have been used in human nutrition for thousands of years, particularly by Asian populations (Arasaki and Arasaki 1985). The globalized economy, the migratory flows observed in the 20th century and the development of mass tourism have made it easier for Western populations to access other culinary practices and foods, whether they are fruits, vegetables or grains that are unusual or even unknown in Western countries.

Algae are some of these new or novel foods (EC 1997) that have benefited from the economic and cultural openness that has followed the globalization of economic exchanges. It is ironic, however, that there is no food more traditional than algae and that the concept of novel food does not stand up to the historical record of their use in human nutrition.

This discrepancy in the perception of algae as a food is the result of very different culinary traditions among populations, but depends above all on the relationship they establish with the sea as a preferred or non-preferred source of food resources.

This book provides an overview of the use of algae (microalgae and macroalgae) and their value mainly in human nutrition and, to a lesser extent, in animal nutrition. Marine microalgae are not widely used in human nutrition compared to macroalgae, which have been eaten in Asia for centuries.

In this context, the overview focuses mainly on macroalgae, which are traditionally used for food applications. However, marine microalgae, when used as food supplements, are also included. Conventionally, the term "algae" refers to macroalgae, with microscopic algae always being referred to as microalgae.

Production methods, namely algae harvesting and cultivation, are described in a global context dominated by Asia, the main region producing algae for direct human consumption (sea vegetables) or indirect consumption (phycocolloids). This assessment is accompanied by a number of prospects for the use of marine algae, particularly for the development of functional foods with limited allergenic potential and the formulation of new foods for fish farming.

These prospects are discussed in light of changing societal demands, which increasingly recognize the need to preserve ecosystems and resources, but which are favorable to the consumption of new products, particularly those derived from the sea (e.g. surimi).

Finally, the book also presents a number of culinary preparations based on algae in order to establish a *continuum* between knowledge and the promotion of this food, which is still largely unknown in Western countries.

References

Arasaki, S. and Arasaki, T. (1985). *Les légumes de mer. Comment être et paraître en forme.* Guy Trédaniel Editions, Paris.

EC (1997). Regulation (EC) No. 258/97 of the European Parliament and of the Council of 27 January 1997 concerning novel foods and novel food ingredients. *Official Journal of the European Communities*, L043, 0001–0006.

1

Algae: Biological and Ecological Characteristics

1.1. The biology of algae

1.1.1. *Definition and taxonomy*

The term "algae" does not correspond to a well-established taxonomic reality; it is used less and less by biologists and has been completely abandoned by phylogeneticists. However, it is still used by naturalists and users of this resource, whether they are harvesters, processors, distributors or consumers. Algae represent a wide variety of photosynthetic organisms that exist in unicellular or multicellular forms. Broadly speaking, a distinction is made between microscopic unicellular algae, or microalgae, and algae that are visible to the naked eye, or macroalgae. Algae consist of eukaryotic cells, which excludes spirulina from this definition, as it is a cyanobacterium with a prokaryotic cell structure. Algae live mainly in aquatic environments, whether marine, brackish or fresh water. Microalgae are present in the aquatic environment but can also be found in particularly humid terrestrial environments such as tree trunks, rocks, glaciers, snowfields and even the walls of certain buildings exposed to the elements. Anecdotally, it should be noted that the purple pigment or B-phycoerythrin secreted by the microalga *Porphyridium cruentum* is believed to be the cause of the reported phenomenon of bleeding statues exhibited in certain particularly humid chapels (personal communication, Gudin Claude). However, another hypothesis based on the involvement of the bacterium *Serratia marcescens*, which produces a brick-red pigment, prodigiosin, is put forward to explain these phenomena (Pitt 1982).

For a color version of all the figures in this chapter, see www.iste.co.uk/fleurence/marinealgae.zip

Microalgae are also present in humid air. This is the case for certain species of Chlorella that have been detected in the atmosphere of the Indian city of Varanasi (Sharma et al. 2006).

Regardless of this morphological distinction between macroalgae and microalgae, it is quite difficult to define algae as a homogeneous plant group from an evolutionary point of view.

The phylogenetic approach has highlighted the evolutionary heterogeneity of algae by distinguishing two taxonomic groups or clades, Rhodobiontes and Chlorobiontes, derived from the same evolutionary sublineage, namely Metabiontes, linked to the green lineage (Lecointre and Le Guyader 2001). From a practical point of view, these two clades include species commonly known as "red algae" (*Palmaria palmata, Chondrus crispus*) and "green algae" (*Ulva lactuca, Caulerpa taxifolia*). These two clades share derived characteristics, in other words characteristics inherited from evolution, such as the absence of a peptidoglycan wall at the chloroplast level, the presence of numerous small DNA nucleoids throughout the chloroplast and the ability to form multicellular organisms.

Brown algae, another taxonomic group of algae, have been grouped within the clade Stramenopiles. This clade contains most of the species used for human consumption, such as *Laminaria digitata* (Kombu), *Fucus vesiculosus* (Bladderwrack) and *Himanthalia elongata* (sea beans) (Lecointre and Le Guyader 2001).

Alongside this phylogenetic classification, some authors take a more traditional approach to classifying algae. Thus, green algae are grouped in the phylum Chlorophyta or Charophyta, red algae in Rhodophyta and brown algae in Ochrophyta (Dawes 2016).

The Charophyta phylum consists of the Charophyceae class. These green algae, also known as Charales, are not used for human consumption. Green algae in the Chlorophyta phylum and Chlorophyceae class, on the other hand, are used as food algae. The Rhodophyta and Ochrophyta phyla consist of the Rhodophyceae (red algae) and Phaeophyceae (brown algae) classes, respectively (Dawes 2016). These two classes of algae include many species (*Palmaria palmata, Porphyra tenera, Laminaria japonica, Undaria pinnatifida*) used in human food as sea vegetables.

However, the identification of biological, genetic or molecular characteristics useful for understanding the evolutionary relationship between algae has not changed the biological definition primarily associated with them.

Microalgae, also known as microphytes, are single-celled photosynthetic microorganisms. Depending on the species and conditions, they can, however, assemble to form undifferentiated multicellular aggregates. Some are equipped with flagella for movement, as is the case with algae of the genus *Chlamydomonas* (see Figure 1.1). Others, such as Chlorella, do not have this type of organelle (see Figure 1.2). The microalgae of the two genera mentioned above are freshwater species. Only species of the genus *Chlorella* are currently used as food supplements.

In this area of use, a marine species belonging to the diatom family is currently attracting particular interest. This is the species *Odontella aurita*, which, like *Chlamydomonas reinhardtii*, has a flagellar system for movement (see Figure 1.3).

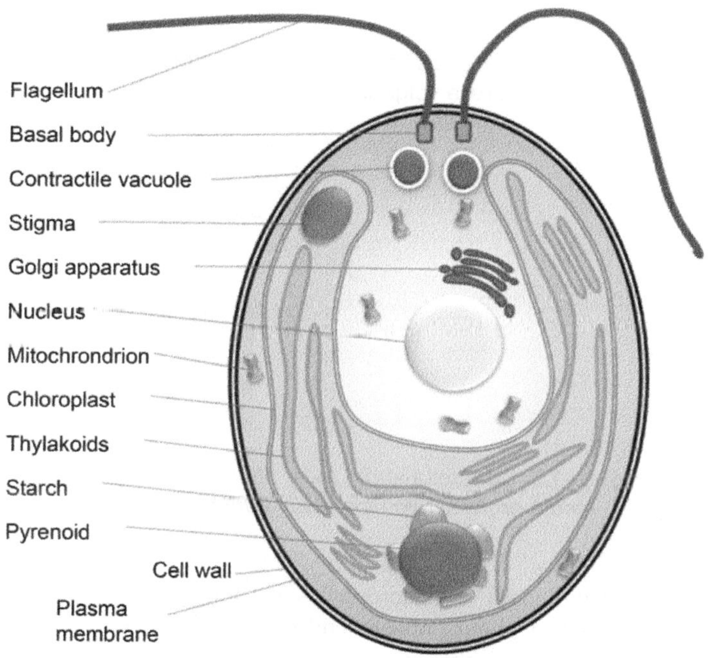

Figure 1.1. *Cellular organization of the microalga* Chlamydomonas reinhardtii *(Fleurence 2021)*

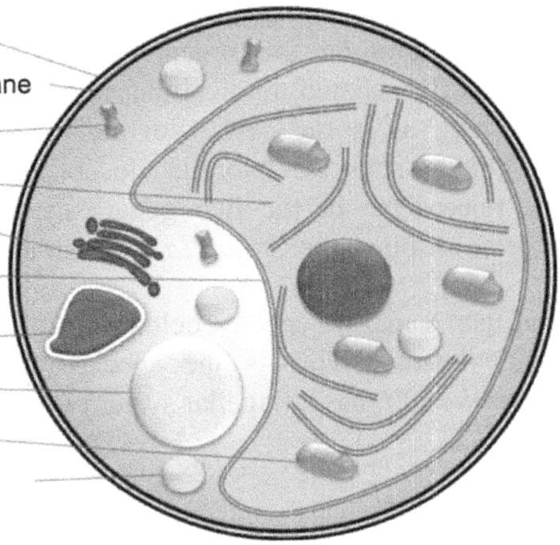

Figure 1.2. *Cellular organization of the microalga* Chlorella vulgaris *(Fleurence 2021)*

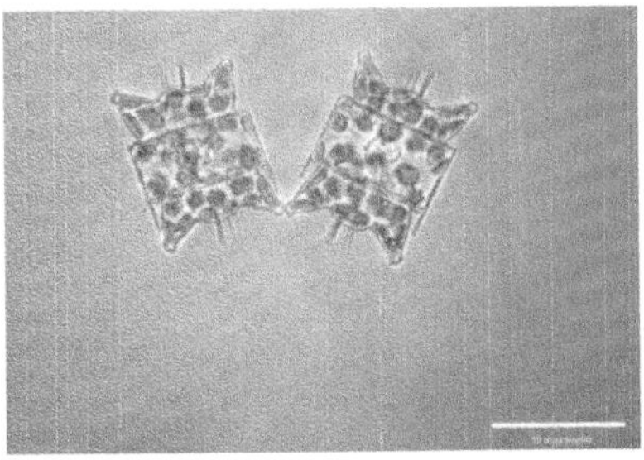

Figure 1.3. Odontella aurita *(see: https://commons.wikimedia.org/wiki/File:Odontella_aurita_cells.jpg)*

Macroalgae, also known as macrophytes or simply "algae", are defined as aquatic plants capable of photosynthesis and those which have a non-vascularized vegetative system called a "thallus" (see Figure 1.4). They share this type of vegetative structure with fungi, which, unlike algae, are unable to perform photosynthesis because they lack chlorophyll pigments. The presence of a thallus initially led botanists to group algae and fungi together in the Thallophytes (lower plants) group. This group was distinguished from the Cormophytes (higher plants), which included plants with a more complex vegetative system consisting of roots, stems, leaves and conducting vessels. This taxonomic classification based on morphological observations is still used today by naturalists. The thallus of an alga, such as *Laminaria digitata*, consists of three distinct morphological parts (see Figure 1.4):

– the upper part called the frond, which consists of multiple blades. These blades constitute the edible part of the algae and are used for food;

– the middle part, or stipe, which enables the algae to stand upright;

– the front part, or haptera, acts as a holdfast, anchoring the algae to its rocky substrate. This organ serves only to anchor the algae and does not absorb minerals, as roots do in cormophytes.

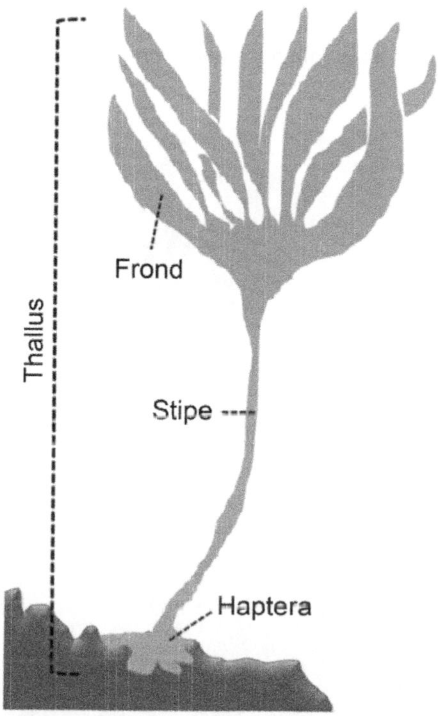

Figure 1.4. *Morphological characteristics of the vegetative apparatus or thallus of the brown alga* Laminaria digitata *(Fleurence 2018)*

Depending on the species, the morphology of the thallus varies greatly, particularly at the stipe level. The latter can be highly developed in brown algae such as *Laminaria digitata* or *Undaria pinnatifida* (see Figure 1.5) or very short in red algae such as *Palmaria palmata* (see Figure 1.6), or even absent in green algae such as *Ulva lactuca* (see Figure 1.7).

Figure 1.5. Undaria pinnatifida *(photo credit © Fleurence J. 2022)*

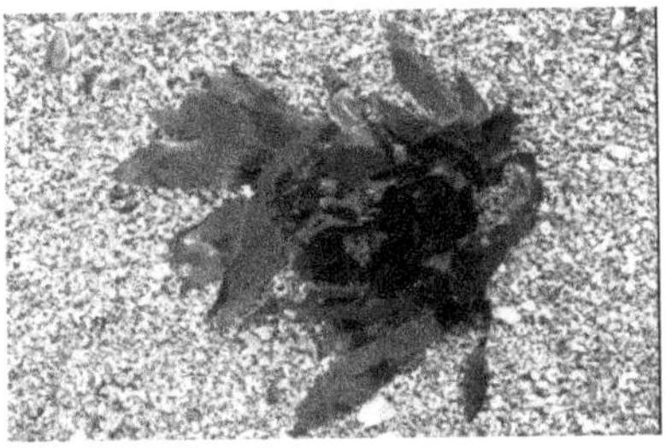

Figure 1.6. Palmaria palmata *(photo credit © Barbaroux O. 2016)*

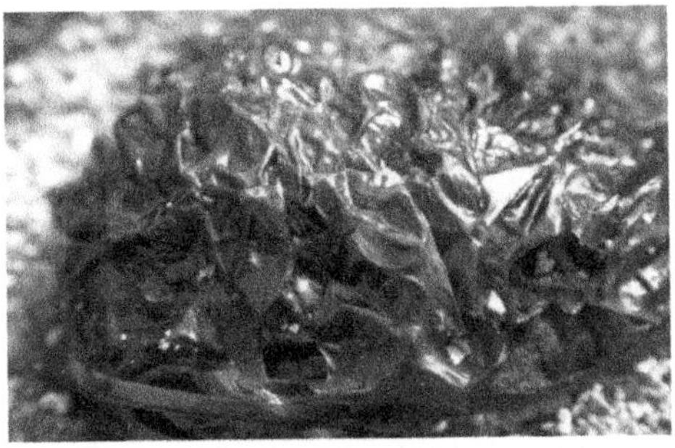

Figure 1.7. Ulva *sp. (photo credit © Barbaroux O. 2016)*

1.1.2. *Development cycles and reproduction*

Algae can reproduce asexually, sexually or even sometimes by cuttings, as is the case with the species *Ulva armoricana*. Knowledge of the life cycle and reproduction of algae is crucial for the development of algaculture in particular. The reproductive cycle of Laminaria (brown algae), and more specifically that of *Laminaria digitata* or *Laminaria japonica*, is currently well understood (see Figure 1.8). During its life cycle, the alga undergoes a heteromorphic digenetic cycle. There is an alternation between a diploid (2n) macroscopic form called sporophyte and a haploid (n) microscopic form called gametophyte. The sporophyte, or asexual form, releases biflagellated, swimming haploid (n) spores from the fertile areas on its frond (sporocysts). These spores develop into microscopic haploid (n) and sexed filaments called gametophytes (n). The gametophytes thus produced release "male" gametes or large "female" gametes called oospheres into the environment. After fertilization and production of the diploid zygote (2n), the latter germinates to produce a seedling or young sporophyte (2n), which will later develop into a mature sporophyte (2n).

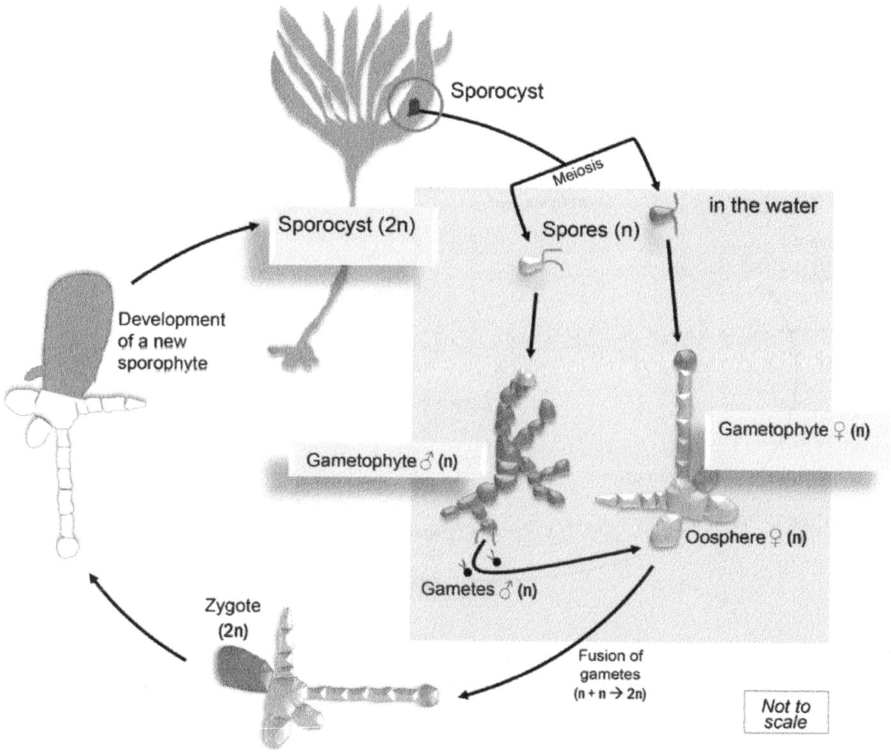

Figure 1.8. *Development and reproduction cycle of brown algae* Laminaria digitata *(Fleurence 2018)*

The life cycle of the brown alga *Undaria pinnatifida* (see Figure 1.9) is also well known and demonstrates many similarities with that of the Laminaria. Like *Laminaria digitata*, this species belongs to the order Laminariales. Its life cycle is, as above, that of a heteromorphic digenetic organism.

However, there are a few differences. In *Undaria* sp., the fertile spore-producing zone is not located at the level of the frond, but at the base of the stipe. This zone is also referred to as the "fertile base" by algae farmers and is characterized by a bulge at the base of the stipe (see Figures 1.4 and 1.5). These fertile bases, known as "Mekabu", are used in human food, while the frond of the algae is marketed under the name "Wakame".

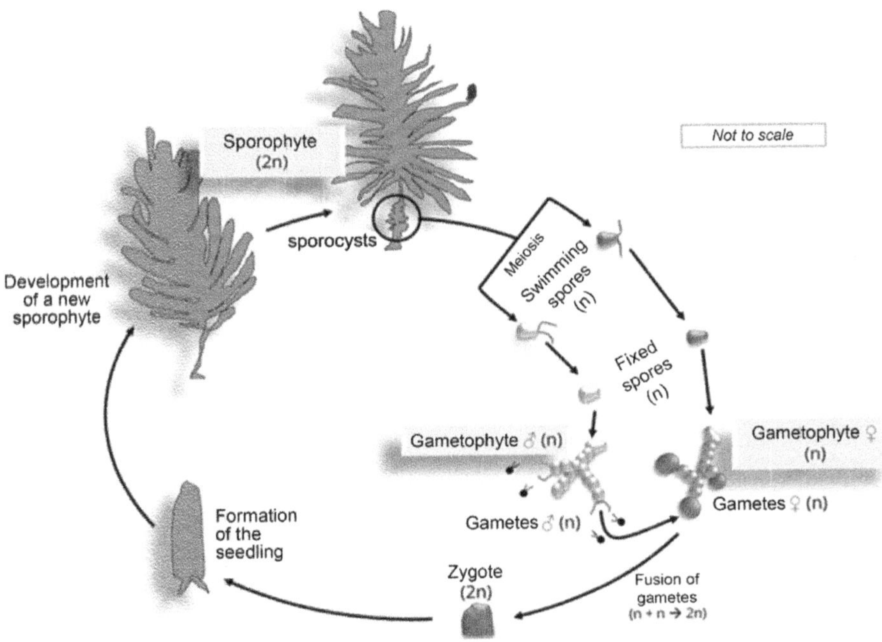

Figure 1.9. *Development and reproduction cycle of brown algae* Undaria pinnatifida *(Wakame) (Fleurence 2018)*

However, other algae species used in human food have very different reproductive cycles. This is particularly true of Ulva, which has an isomorphous digenetic cycle. The diploid sporophyte forms (2n) and haploid gametophyte forms (n) are macroscopic and morphologically similar (see Figure 1.10). This characteristic is a disadvantage in controlling Ulva cultivation, as it is very difficult to distinguish the sporophyte from the gametophyte, unlike Laminariales (*Laminaria* sp., *Undaria* sp.), where the two stages of development are identifiable and easily isolable.

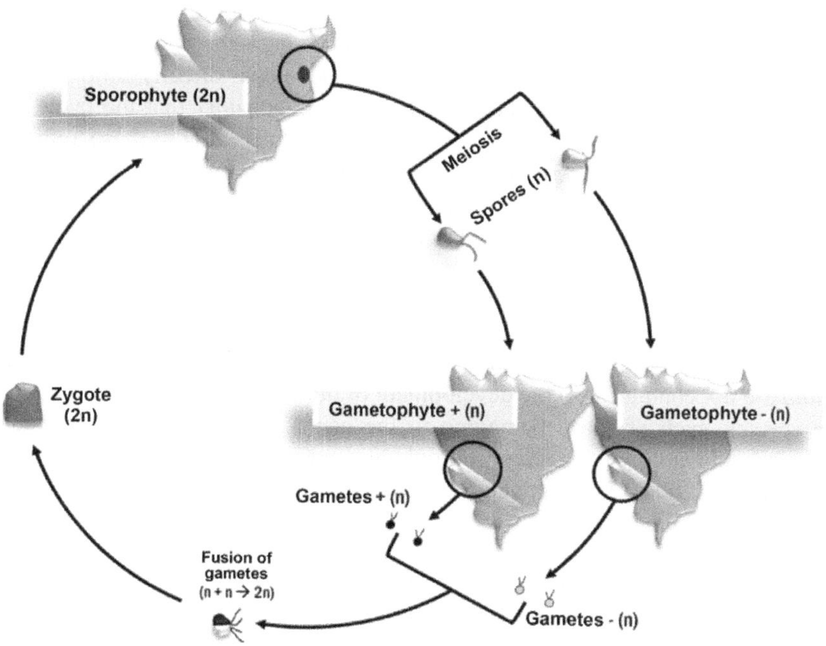

Figure 1.10. *Development and reproduction cycle of the green alga* Ulva *sp. (sea lettuce) (Fleurence 2018)*

In addition to a reproductive cycle based on an asexual pathway (spores) and a sexual pathway (gametes), some species can also reproduce by fragmentation of the thallus. This process of "cutting" is observed in certain Ulvales, such as *Ulva. armoricana* or *Ulva prolifera.* When these species detach from their rocky substrate and become pelagic, they are capable, under certain conditions of temperature and nitrogen nutrient concentrations, of generating new individuals from fragments of their thalli. Thus, the floating biomass of *U. prolifera* can increase by 6.30% per day with only 10 μM of dissolved inorganic nitrogen and a water temperature of 20°C (Keesing et al. 2011). This mechanism is one of the reasons for the proliferative, even invasive, nature of these algae and the onset of green tides.

Other species, such as the red alga *Palmaria palmata,* have a biological cycle that has long been misunderstood. The life cycle of this alga was initially described under in vitro culture conditions by van der Meer and Todd in 1980 (Deniaud 2002). It is described as digenetic and based on the alternation of two generations of individuals. The algae are initially described in a macroscopic diploid (2n) or tetrasporophyte form, producing haploid tetraspores (n) through meiosis, half of which germinate into male gametophytes (n) and the other half into female gametophytes (n). The male gametophytes (n) are macroscopic and morphologically similar to the tetrasporophytes (2n), while the female gametophytes are microscopic. The zygote (2n) resulting from fertilization develops directly into a diploid tetrasporophyte (Cabioc'h et al. 1992). Other authors consider that the cycle of this species is more complex, however, as it is trigenetic. According to these authors, the cycle is based on the succession of three generations during the development of the alga: tetrasporophytes (2n), gametophytes (n) and carposporophytes (2n). The carposporophyte, a diploid individual derived from the development of the zygote, produces diploid spores that generate new diploid tetrasporophytes (De Reviers 2002).

These differences in interpreting the development cycle of the algae *Palmaria palmata* are not conducive to the development of this species' cultivation. In other red algae species, such as those belonging to the genus *Gracilaria,* the presence of a trigenetic cycle involving a carposporophyte generation is, however, well known (Cabioc'h et al. 2006). The cultivation of Gracilaria, particularly for agar production, is therefore common practice.

Knowledge of the biological cycle of algae and, consequently, control of this cycle is one of the conditions for successful algae cultivation. The algacultural development of Laminariales (*Laminaria* sp., *Undaria* sp.) or Gracilariales around the world is a direct result of this observation.

1.1.3. *Pigment composition*

Algae, which have the distinctive feature of living under a layer of water that limits the penetration of light, have developed unique pigments that distinguish some of them from those found in terrestrial plants.

This pigment composition is the basis for their taxonomic classification into green algae (Chlorophyceae), red algae (Rhodophyceae) and brown algae (Phaeophyceae). As seen above, this traditional classification is still used by biologists and is widely used by all actors in the algae industry. While all algae contain chlorophyll a, they differ from one another in their composition of additional pigments known as supernumerary pigments (Table 1.1).

Chlorophyll a, a photosynthetic pigment common to all species of algae and all photosynthetic plants, is characterized by two absorption maxima located at 432 and

662 nm, respectively (Dumay and Morançais 2016). Chlorophylls b–d have fairly similar chemical structures, but their chromophores are different enough to explain the differences in spectral properties at the light absorption maxima (see Figures 1.11 and 1.12 and Table 1.2). The supernumerary pigments have maximum absorption wavelengths that differ from those of chlorophyll a, and these differ depending on the nature of the pigments (Table 1.2).

Botanical class	Main pigment	Supernumerary pigment
Chlorophyceae (green algae)	Chlorophyll a	Chlorophyll b
Phaeophyceae (brown algae)	Chlorophyll a	Chlorophyll c and an excess of carotenoids (carotene, xanthophyll)
Rhodophyceae (red algae)	Chlorophyll a	Phycoerythrin Phycoerythrocyanin Allophycocyanin Chlorophyll d

Table 1.1. *Pigment composition of different classes of algae (Fleurence 2018)*

Botanical class	Supernumerary pigment (color)	λ_{max} absorption (nm)
Chlorophyceae (green algae)	Chlorophyll b (olive green)	457
Pheophyceae (brown algae)	α-carotene (yellow) β-carotene (orange-yellow) Chlorophyll c (light green) Fucoxanthin (orange)	447/475 454/484 446/578/627 446/468
Rhodophyceae (red algae)	Phycoerythrin e (purple) Allophycocyanin (blue) Phycoerythrocyanin (orange) Chlorophyll d	540/565 618/671 568/570 401/455/696

Table 1.2. *Spectral characteristics of some supernumerary pigments used in the classification of macroalgae (Dumay and Morancais 2016)*

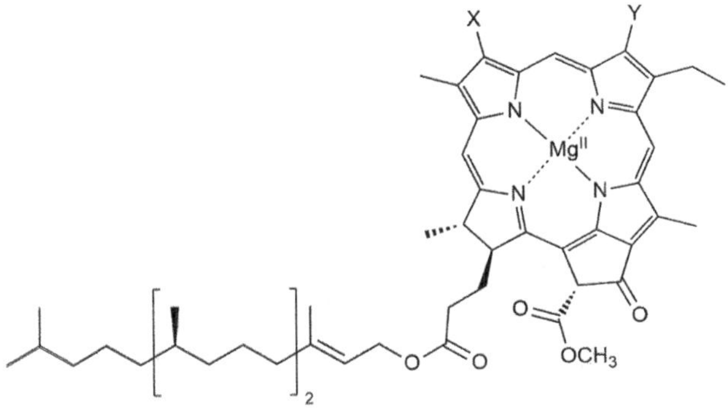

chl a : X: $CH{=}CH_2$ Y: CH_3
chl b : X: $CH{=}CH_2$ Y: CHO
chl d : X: CHO Y: CH_3

Figure 1.11. *Structure of the chromophores of chlorophylls a, b and d (Fleurence 2018)*

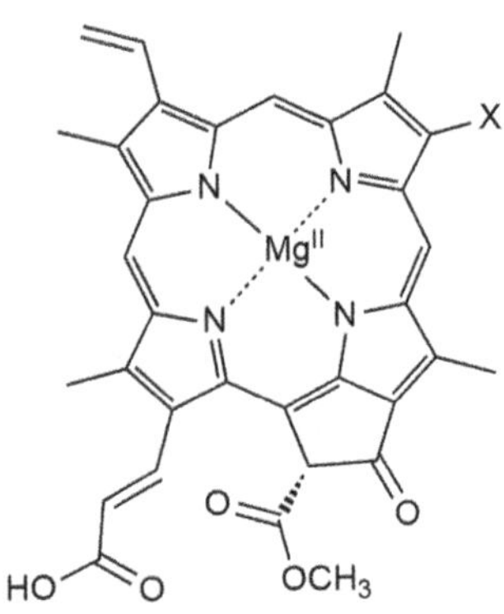

chl c1 : X: CH_2-CH_3
chl c2 : X: $CH{=}CH_2$

Figure 1.12. *Structure of the chromophores of chlorophylls c1 and c2 (Fleurence 2018)*

The red, purple and sometimes pink color of red algae is mainly due to phycobilins, chromophores that are constituents of phycobiliproteins and, more specifically, phycoerythrin. These pigments complement the action of chlorophyll a and enable red algae to grow at depths ranging from the intertidal zone (the area affected by tides) to the subtidal zone (the area never uncovered by the tides, including spring tides). Phycobiliprotein pigments are highly unique molecules resulting from the combination of a chromophore, an open tetrapyrrole nucleus, with a protein (see Figure 1.13). In the case of phycoerythrin, this chromophore or phycoerythrobilin restores the absorbed light energy in the form of fluorescence at a specific wavelength ((λ_{max}) fluorescence emission 578 nm for R-phycoerythrin).

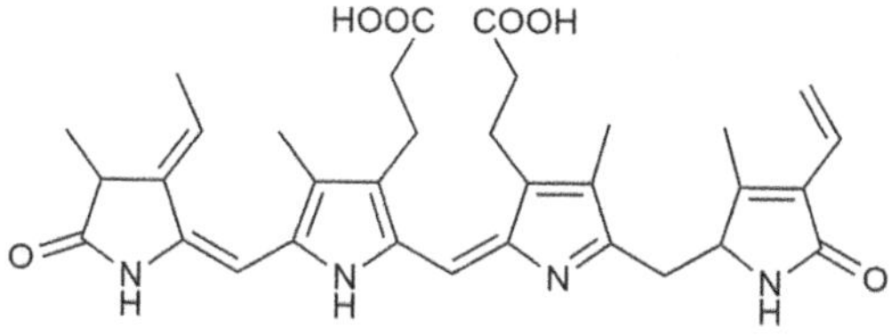

Figure 1.13. *Structure of the chromophore or phycoerythrobilin of phycoerythrin (Fleurence 2018)*

The phycoerythrin and phycocyanin family, original and characteristic pigments of red algae, and even cyanobacteria in the case of C-phycocyanin, is of interest for agri-food applications, particularly as a natural colorant. This pigment diversity, which is responsible for the definition of the different botanical classes of algae, is also responsible for the diversity of colors associated with these marine organisms.

Thus, in red algae or Rhodophyceae, a whole range of hues can be observed, from light pink to dark red. This color variation is the result of the presence of other pigments such as phycocyanin alongside the supernumerary red pigment phycoerythrin (see Figure 1.14).

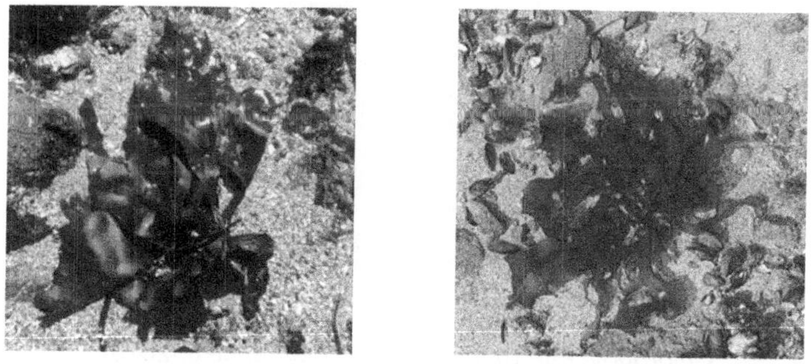

a) b)

Figure 1.14. *Phenotypic variation in the red color of Rhodophyceae according to species (a)* Palmaria palmata, *b)* Grateloupia turuturu*)* *(photo credit © Fleurence J. 2022)*

Brown algae (Pheophyceae) have a thallus ranging from dark brown to yellow-brown with orange hues depending on the concentration of carotenoid pigments (see Figure 1.15).

a)

b)

Figure 1.15. *Phenotypic variation in the brown color of Phaeophyceae according to species (a)* Undaria pinnatifida, *b)* Fucus vesiculosus*)* *(photo credit © Fleurence J. 2022)*

Finally, green algae, or Chlorophyceae, due to the simultaneous and predominant presence of chlorophyll pigments of types a and b, have a green thallus (see Figure 1.16).

Figure 1.16. Ulva *sp., an alga belonging to the class Chlorophyceae (photo credit © Fleurence J. 2023)*

However, algae color can be misleading, as it can vary depending on environmental conditions for certain species. This criterion can vary depending on whether the algae are submerged under a layer of water of varying depth or exposed on the foreshore (Box 1.1). Other parameters such as the conditions in which the algae develop or are produced (growth in the natural environment or in culture) can also have an impact on the color of the thallus of certain algae. This is the case for red algae, and more particularly for cultivated species belonging to the genus *Porphyra*, whose colors can change from green to brown to red depending on the growing conditions (De Reviers 2002).

In addition to this example, some species classified in the Phaeophyceae family (brown algae) may also have a dark green thallus rather than a brown thallus. This is particularly the case for the species *Himanthalia elongata*, which is valued in human food as a sea vegetable.

Recognition of botanical families of algae on the foreshore. Taxonomic identification

a) b)

Figure 1.17. Fucus *sp. observed under a layer of water (a) and in the open air on the foreshore (b) (photo credit © Fleurence J. 2022)*

The presence or absence of a layer of water during observation may cause variations in color assessment depending on the observer. As a general rule, color should not be used as a taxonomic identification criterion for organisms that have emerged or are emerging from the water.

Box 1.1. *Taxonomic recognition methodology of algae based on color*

1.1.4. *Size*

In addition to color, algae also vary greatly in size, depending on the species and stage of development. A species such as the brown algae *Laminaria japonica*

(*Saccharina japonica*), measures on average 4–6 m in length and 30 cm in width. However, the thallus of the latter can reach 10 m when cultivated (Perez 1997). In contrast, the red alga *Palmaria palmata* has a thallus that rarely exceeds 30 cm in size.

1.2. Ecology of algae

Algae are aquatic plants that are autotrophic for carbon and depend on sunlight for photosynthesis. Their bathymetric distribution is therefore limited by this physiological constraint.

Three bathymetric zones are identified based on light penetration:

– the euphotic zone, where light penetration is sufficient for photosynthesis to occur; this zone can extend to a depth of 200 m;

– the oligophotic zone, where light still penetrates but in insufficient quantities for photosynthesis to occur;

– the aphotic zone, where no light penetrates; this zone corresponds to the abyssal depths.

In terms of the quality of the visible light spectrum (400–700 nm), red radiation is largely blocked at a depth of 10 m, and only blue and green radiation reaches the lower layers of the euphotic zone (< 100 m). This light filtering is responsible for the distribution of algae according to their pigment content. For example, green algae, which, like land plants, only have chlorophylls a and b, are able to absorb red and indigo light, reflecting the rest of the light spectrum, including green, which is responsible for their color. They are found in water layers where red radiation can still penetrate and are therefore located at shallower depths than brown or red algae (see Figure 1.18).

More generally, most algae species used and valued for human consumption are found in the tidal zone. This zone, also known as the foreshore, is accessible on foot and remains the harvesting area for harvesters or seaweed gatherers (see Figure 1.19). Other species, particularly brown algae belonging to the genus *Laminaria*, are found at the lower limit of the intertidal zone and at the upper limit of the subtidal zone, which is never exposed, even during spring tides.

Laminaria are therefore only partially accessible on foot during very high tides, thus limiting their harvest. In Brittany, France, the seaweed field ends at a depth of 30 m, which therefore represents the ecological limit for seaweed harvesting on the Atlantic coast (Quéro and Vayne 1998).

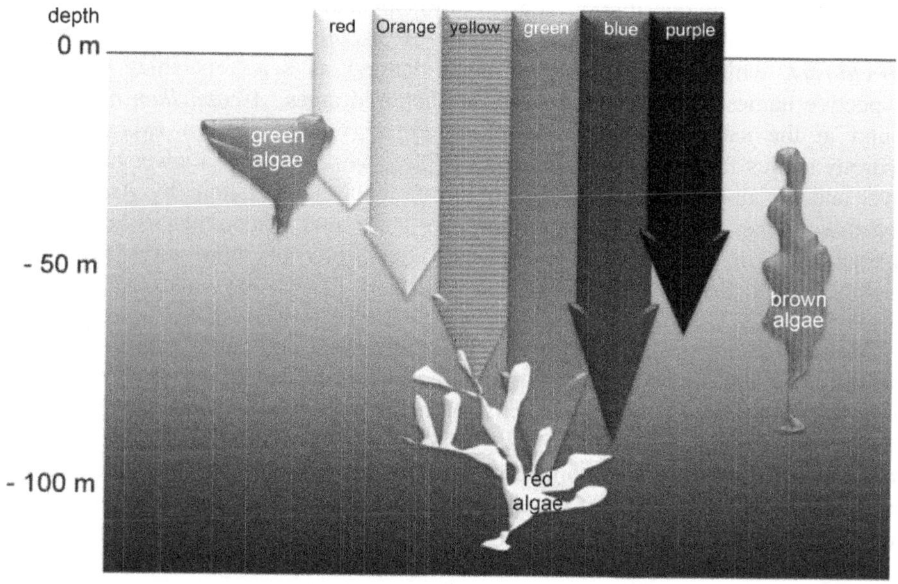

Figure 1.18. *Evolution of light radiation in the visible spectrum as a function of bathymetry (Fleurence 2018)*

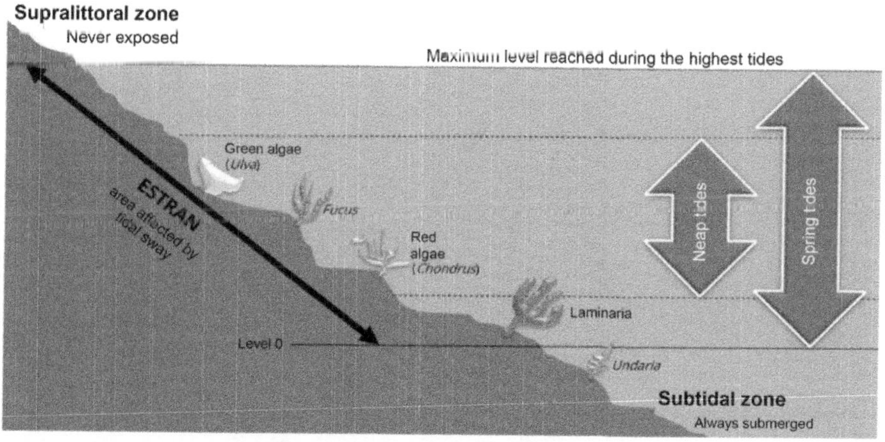

Figure 1.19. *Example of bathymetric distribution of green, red or brown algae used in human food (Fleurence 2018)*

The bathymetric distribution of algal flora can also be influenced by waves. This is particularly observed for species such as *Ascophyllum nodosum* and *Fucus vesiculosus*, which are two brown algae valued as sea vegetables under the respective names of Wrack and Kelp. In sheltered areas, *Ascophyllum nodosum* is found at the same bathymetric level as *Fucus vesiculosus*. In exposed areas, hydrodynamics favor the establishment of *Fucus vesiculosus* at a lower bathymetric level and are not conducive to the colonization of the foreshore by *Ascophyllum nodosum*. This has led some authors to state that *Ascophyllum nodosum* is characteristic of calm waters and *Fucus vesiculosus* of more turbulent waters (Cabioc'h et al. 1992).

Microalgae are present in many aquatic, terrestrial and aerial environments. In the air, they are found in the humid atmosphere thanks to the phenomenon of aerosolization (Box 1.2). In aquatic environments, particularly marine environments, marine microalgae can live in pelagic form in the water column. They then constitute the phytoplankton compartment. They can also develop a benthic lifestyle. In this case, they form part of the microphytobenthos, which combines microalgae and cyanobacteria. The microphytobenthos in the intertidal zone generates a brown-greenish biofilm that is characteristic of mudflats (see Figure 1.21).

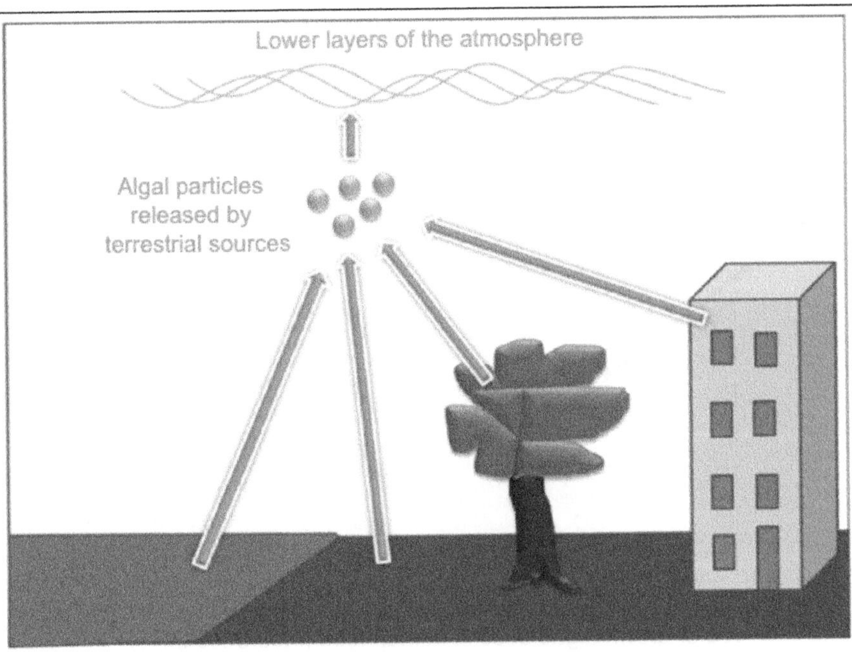

Figure 1.20. *Ecosystems responsible for the release of microalgae in the lower layers of the atmosphere (Fleurence 2021)*

Microalgae are transferred to the lower atmosphere via a process called aerosolization, which transforms them into fine algal particles. They are released from their original environment by wind and upward air currents.

Box 1.2. *Microalgae aerosolization phenomenon*

Figure 1.21. *Biofilm forming the colonization of the bottom of a mudflat by microphytobenthos and cyanobacteria (photo credit © Rosa P. 2022)*

All these biological and ecological characteristics of algae and microalgae are essential elements of knowledge about the resource that are necessary for understanding the choice of species exploited and used in human food.

1.3. References

Cabioc'h, J., Floc'h, JY., Le Toquin, A., Boudouresque, CF., Meinez, A., Verlaque, M. (1992). *Guide des algues des mers d'Europe*. Delachaux et Niestlé, Neuchâtel-Paris.

Cabioc'h, J., Floc'h, JY., Le Toquin, A., Boudouresque, CF., Meinez, A., Verlaque, M. (2006). *Guide des algues des mers d'Europe*. Delachaux et Niestlé, Neuchâtel-Paris

Dawes, C. (2016). Macroalgae systematics. In *Seaweed in Health and Disease Prevention*, Fleurence, J. and Levine, I. (eds). Academic Press, Elsevier, Amsterdam.

Deniaud, E. (2002). Structure et interaction des polymères dans les parois cellulaires de *Palmaria palmata* (Rhodophyta). PhD Thesis, Université Pierre Marie Curie, Paris.

De Reviers, B. (2002). *Biologie et phylogénie des algues 1*. Belin, Paris.

Dumay, J. and Morançais, M. (2016). Proteins and pigments. In *Seaweed in Health and Disease Prevention*, Fleurence, J. and Levine, I. (eds). Academic Press, Elsevier, Amsterdam.

Fleurence, J. (2018). *Les algues alimentaires : bilan et perspectives*. Lavoisier Tec & Doc., Paris.

Fleurence, J. (2021). *Microalgae: From Future Food to Cellular Factory*. ISTE Ltd, London and John Wiley & Sons, New York.

Keesing, J.K., Dongyan, L., Fearns, P., Garcia, R. (2011). Inter-and intra-annual patterns of *Ulva prolifera* green tides in the Yellow Sea 2007-2009, their origin and relationship to the expansion of coastal seaweed aquaculture in China. *Marine Pollution Bulletin*, 62, 1169–1182.

Lecointre, G. and Le Guyader, H. (2001). *Classification phylogénétique du vivant*. Belin, Paris.

van der Meer, J.P. and Todd, E.R. (1980). The life history of Palmaria palmata in culture. A new type for the Rhodophyta. *Canadian Journal of Botany*, 58(11), 1250–1256.

Perez, R. (1997). *Ces algues qui nous entourent*. Ifremer, Paris.

Pitt, T.L. (1982). State of the art: Typing of *Serratia marcescens*. *Journal of Hospital Infection*, 3, 9–14.

Quéro, J.C. and Vayne, J.J. (1998). *Les fruits de la mer et plantes marines des pêches françaises. Les encyclopédies du naturaliste*. Delachaux et Niestlé, Ifremer, Lausanne, Paris.

Sharma, N.K., Singh, S., Rai, A.K. (2006). Diversity and seasonal variation of algal particles in the atmosphere of subtropical city in India. *Environmental Research*, 102, 252–259.

2

Algae Species Used in Human Nutrition

2.1. Phaeophyceae (brown algae)

Asian populations have been consuming macroalgae since the ancient times. Fragments of algae have been identified in fossilized meal remains at Japanese and Korean archaeological sites, and this practice is estimated to date back 10,000 years (Nisizawa et al. 1987; Fleurence 2016). Other sources indicate more recent use dating back to the 4th century CE in Japan and the 6th century in China (FAO 2004).

The species involved in this highly traditional consumption belong to the genera *Sargassum*, *Eisenia*, *Hizikia* and *Undaria*, exclusively brown algae, some of which are still used today for human consumption (Chapter 6, section 6.2).

Phaeophyceae currently remain the main algal resource used in human nutrition, either directly (sea vegetables) or indirectly (phycocolloids or food additives). Among the latter, species belonging to the genus *Laminaria* are mainly exploited on a large scale. At the end of the 1990s, the annual harvest of Laminaria was approximately 683,000 tons in dry weight (3.4 10^6 tons in fresh weight) (Zemke-White and Ohno 1999; Fleurence 2016) (Table 2.1). Among the Laminaria species, the most widely used is *Laminaria japonica* (*Saccharina japonica*), mainly known as Kombu. This species is found on the Japanese coast of the Hokkaido region and on the northeast Pacific coast of Japan. It is also present on the coasts of China and South Korea. Over the last decade, the production of *Laminaria japonica* (*Saccharina japonica*), the main Laminaria used for food applications, has grown steadily. At the end of the 2010s, it

For a color version of all the figures in this chapter, see www.iste.co.uk/fleurence/marinealgae.zip

reached nearly 12 million tons of fresh weight (see Figure 2.1). Ninety percent of this production is carried out by China, which harvests this alga through algaculture (see Figure 2.2).

Genus	World production (tons of dry weight)
Laminaria sp.	682
Undaria sp.	101,708
Hizikia sp.	7,497
Sargassum sp.	7,649
Fucus sp.	82

Table 2.1. *Global production of selected species used in human food (Zemke-White and Ohno 1999; Fleurence 2016)*

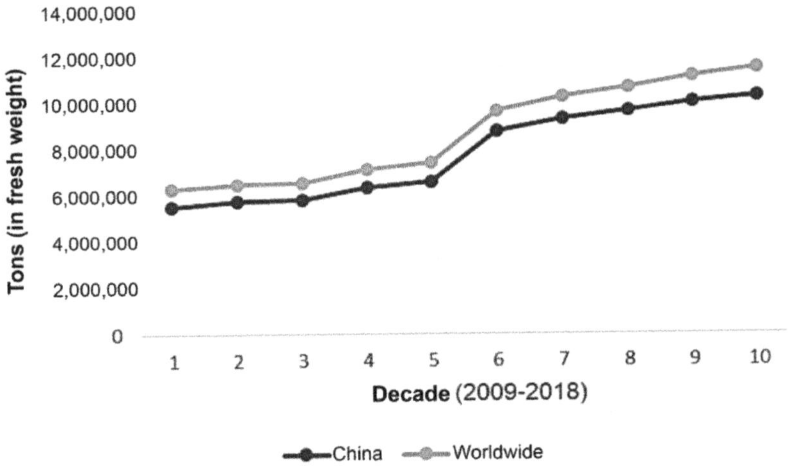

Figure 2.1. *Evolution of the production of* Laminaria japonica *(*Saccharina japonica*) worldwide and in China between 2009 and 2018 inclusive (1. 2009, 2. 2010, 3. 2011, 4. 2012, 5. 2013, 6. 2014, 7. 2015, 8. 2016, 9. 2017, 10. 2018) (FAO 2020; Fleurence 2024)*

Other species of Laminaria, such as *L. angustata*, *L. religiosa*, *L. diabolica* and *L. groenlandica,* are also harvested for human consumption (Nisizawa et al. 1987; Zemke-White and Ohno 1999). In France, the species *Laminaria digitata* is mainly exploited in northern Brittany for these two types of food use (sea vegetables, food additives). It is marketed under the name "Breton Kombu" to distinguish it from *Laminaria japonica*. However, this distinction in the common name of the two species is often not respected.

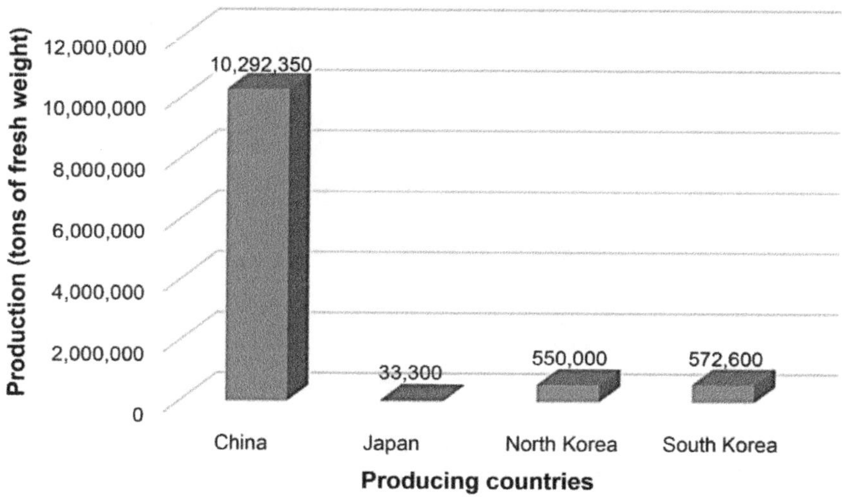

Figure 2.2. *Global distribution of* Laminaria japonica *(*Saccharina japonica*) production (in tons of fresh weight) by algaculture (FAO 2020; Fleurence 2024)*

The genus *Undaria*, through the species *Undaria pinnatifida*, is also highly valued, mainly as a sea vegetable under the name Wakame (Table 2.1). This species, native to the coasts of Japan and South Korea, was accidentally introduced with imported oyster spat in the Thau lagoon (southern France) in the early 1970s (Perez et al. 1981) (see Figures 2.3 and 2.4).

Its cultivation was developed by Ifremer off the coast of Ouessant in the 1990s, and this species is now harvested as a sea vegetable (Chapter 6, section 6.2). The cycle of this alga introduced to our coasts shows notable differences in terms of seasonality in its development. Indeed, the algae present in the Thau lagoon reaches optimal frond development, the part used for human consumption, during the spring and more specifically in March and April (see Figure 2.5). This is not the case for *Undaria pinnatifida* cultivated on the Atlantic coast of Ouessant, where thallus development reaches maturity in February (Perez et al. 1984). Since *Undaria pinnatifida* was first cultivated, other production sites have sprung up on the Atlantic coast (Oléron) and the English Channel (Saint-Malo).

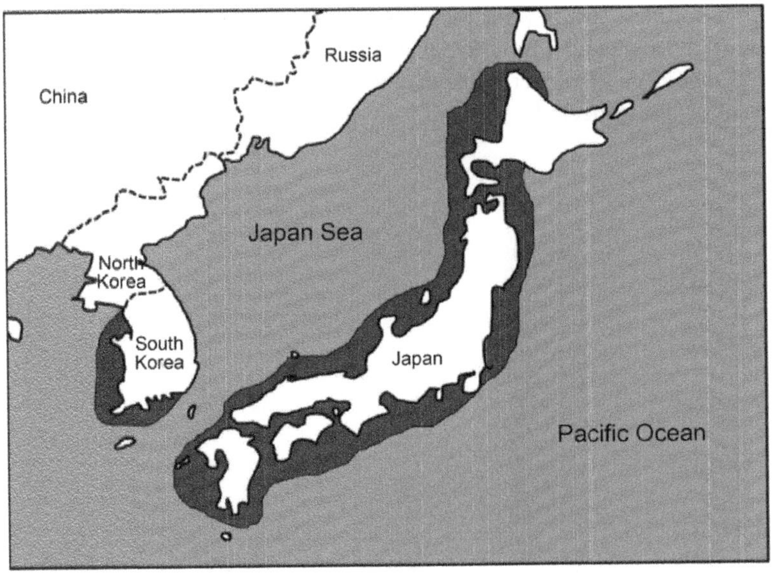

Figure 2.3. *Geographical distribution of the seaweed* Undaria pinnatifida *(Wakame) on the coasts of Japan and Korea (Fleurence 2018)*

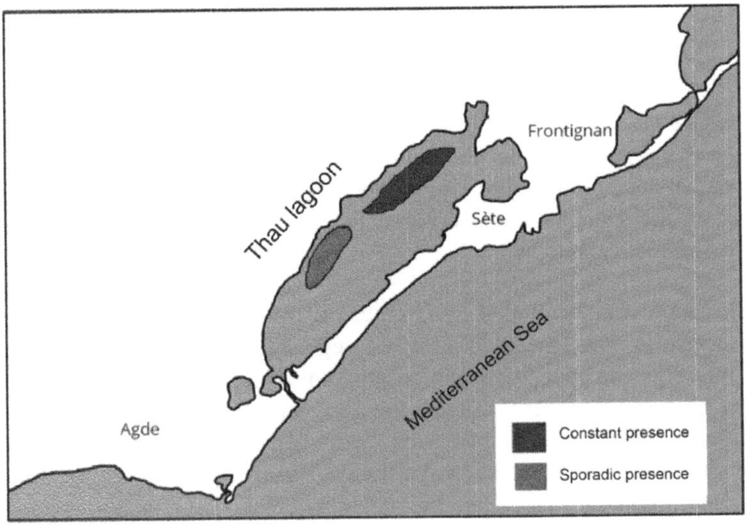

Figure 2.4. *Geographic distribution of the introduced alga* Undaria pinnatifida *(Wakame) in the Thau lagoon (Hérault, France) (Fleurence 2018)*

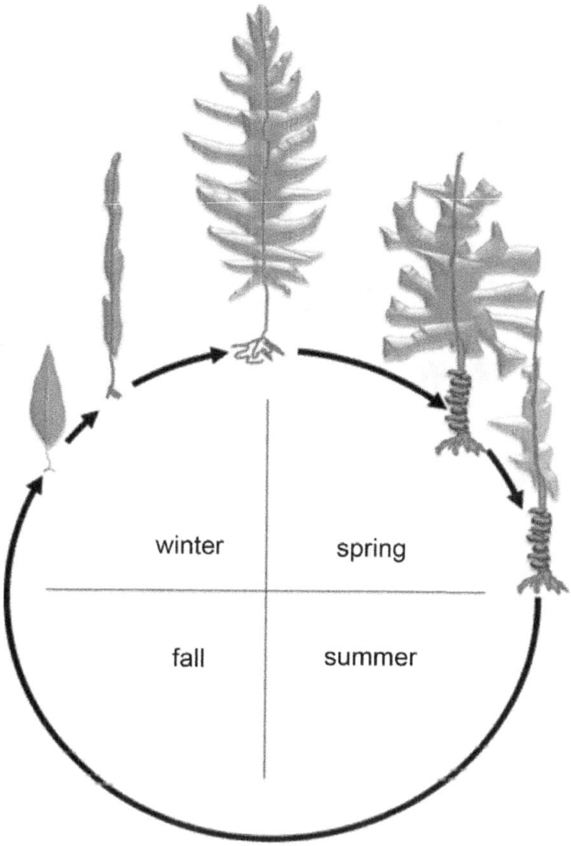

Figure 2.5. *Development cycle of the alga* Undaria pinnatifida *present in the Thau lagoon (Fleurence 2018)*

However, the development of *Undaria pinnatifida* cultivation, an alga highly prized by Asian consumers, is not always well accepted in Europe. This alga, which is not native to European coasts, is considered invasive (Murphy et al. 2016) and its cultivation is seen as an effective means of spreading the species to coasts where it was previously absent. *Undaria* was reported on the Portuguese coast between 1999 and 2007 (Veiga et al. 2014). However, it has mainly been observed on the British, Irish and Scottish coasts (see Figure 2.6) (Minchin and Nunn 2014). Some authors believe that all the conditions are now in place for this species to spread to the Norwegian coast (Minchin and Nunn 2014). This image of an "invasive algae" is affecting the use of this species in human food in some European countries and therefore limiting its commercial development as a sea vegetable.

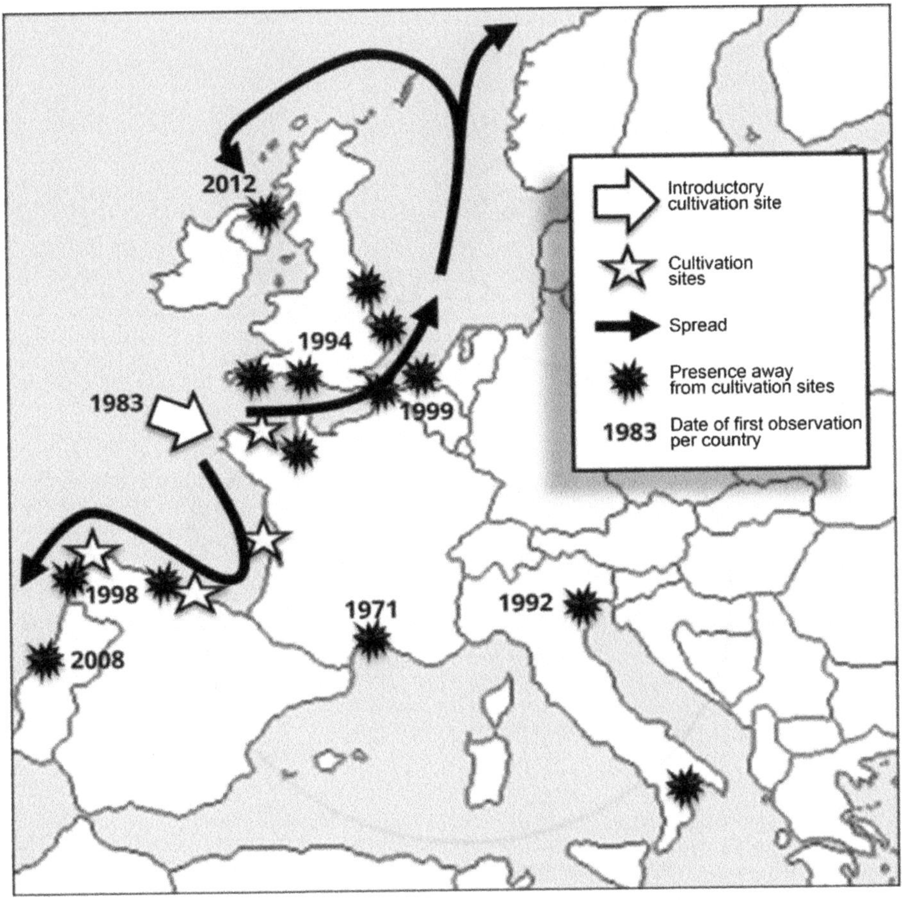

Figure 2.6. *Spread of the alga* Undaria pinnatifida *(Wakame) along European coasts (Fleurence 2018)*

The species *Hizikia fusiformis* (*Sargassum fusiforme*), a brown alga belonging to the Sargassaceae family, is highly prized as a sea vegetable in Japan, where it is sold under many commercial names (Chapter 6, section 6.2) (Nisizawa et al. 1987). This species is found on the Japanese coast of Honshu facing the Pacific Ocean (Ohno and Largo 1998).

Algae of the genus *Sargassum* are also consumed by humans in certain countries such as Bangladesh, Malaysia, Vietnam, Thailand and the Philippines (Zemke-White and Ohno 1999). In Korea, it is the species *Sargassum horneri* that is mainly used as a sea vegetable by local populations.

In Europe, Sargassum are introduced species and are characterized by their high dispersal and colonization capacity in ecosystems. This is particularly true of the species *Sargassum muticum* (see Figure 2.7), which is native to Japan and migrated to European coasts after colonizing the entire North American coast (Deysher and Norton 1981). In France, this alga is present on all coasts and its proliferation can cause numerous problems for human activities, such as fishing and shellfish farming (Belsher and Pommelec 1988). The negative impact of *Sargassum muticum* proliferation on human activities and on the development of other algae through the canopy it develops often leads to it being classified as an invasive species. The invasive and non-native nature of this species currently represents a definite barrier to the development of its use in human food in European countries.

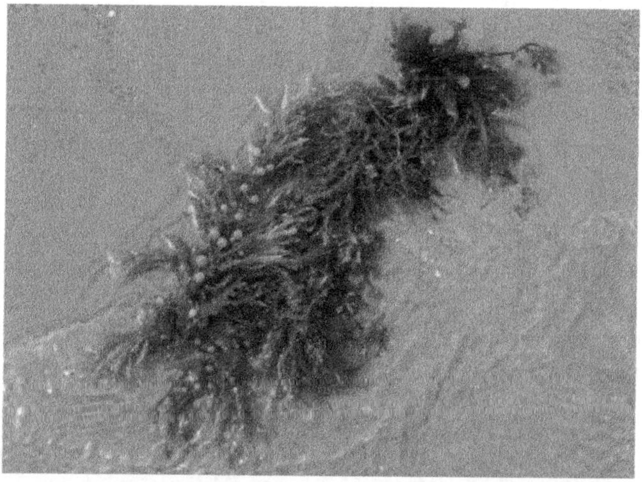

Figure 2.7. Sargassum muticum *(see: https://commons.wikimedia.org/wiki/File:Sargassum_muticum_stranded.jpg)*

Fucales, such as *Fucus vesiculosus*, and *F. serratus*, are also consumed as sea vegetables, but mainly in Europe and more particularly in France. This is also the case for the brown alga *Himanthalia elongata* (Fleurence 2016). This species is found on the European Atlantic coast and is easily obtainable.

2.2. Rhodophyceae (red algae)

Red algae are also used in human food. Algae belonging to the genus *Porphyra* are highly prized as sea vegetables.

The species *Porphyra tenera* (*Pyropia tenera*) and *Porphyra yezoensis* are consumed under the name "Nori". These two species, which occur naturally on the Japanese and Korean coasts, have been cultivated in both countries since the 17th century (FAO 2004).

Global production of *Porphyra* through algaculture, mainly *Porphyra tenera,* was estimated at nearly 131,000 tons of dry product at the end of the 1990s (approximately 655,000 tons of fresh product) (Table 2.2). The exploitation of this alga has grown significantly to reach approximately 1,600,000 tons (fresh product) in the last decade (see Figure 2.8).

On the northern European coast, particularly in Brittany, the species *Porphyra umbilicalis* is mainly found.

Algae belonging to the genus *Gracilaria*, are also consumed as sea vegetables. Approximately 10 different species are used as edible algae around the world (Chapter 6, section 6.2) (Table 2.3). They are also used to produce food additives (agar) (Chapter 6, section 6.7).

The species *Gracilaria verrucosa* is consumed in France, Indonesia, Japan and Korea under the name "Ogonori". Gracilaria are algae with a very wide geographical distribution and are often described as cosmopolitan algae. They are found in both temperate and tropical zones. In Europe, they are present on the Atlantic and Mediterranean coasts.

Genus	World production (tons of dry weight)
Porphyra sp.	130,614
Gracilaria sp.	50,165
Eucheuma sp.	25,545
Gelidium sp.	18,773
Chondrus sp.	12,213
Palmaria sp.	10

Table 2.2. *Global production in 1994 of selected red algae species used in human food (Zemke-White and Ohno 1999; Fleurence 2018)*

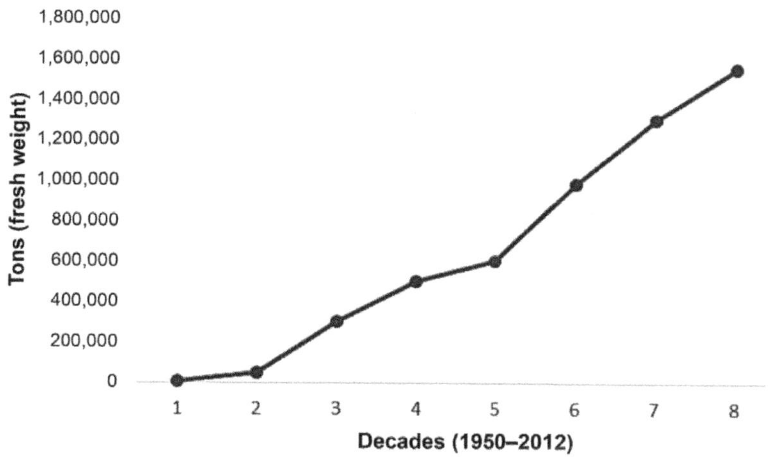

Figure 2.8. *Evolution of the production of the red algae* Porphyra tenera *(Nori) by algaculture, from 1950 to 2012 (1: 1950, 2: 1960, 3: 1970, 4: 1980, 5: 1990, 6: 2000, 7: 2010, 8: 2012) (see: https://www.fao.org/fishery/docs/DOCUMENT/aquaculture/CulturedSpecies/file/en/en_nori.htm)*

Gracilaria species	Country of consumption
Gracilaria asiatica	Vietnam
Gracilaria bursa-pastoris	Japan
Gracilaria domingensis	Brazil, Chile, Caribbean
Gracilaria changii	Thailand
Gracilaria eucheumoides	Indonesia, Vietnam
Gracilaria crassissima	Caribbean
Gracilaria coronopifera	Hawaii, Vietnam
Gracilaria lemaneiformis	Japan
Gracilaria salicornia	Japan, Thailand, Vietnam
Gracilaria verrucosa	Japan, Indonesia, France, Korea

Table 2.3. *Examples of* Gracilaria *species consumed as food worldwide (Zemke-White and Ohno 1999; Fleurence 2018)*

Algae of the genus *Grateloupia* are used in Japan as sea vegetables, as well as agorophytes (Arasaki and Arasaki 1985). They share this dual application with

Gracilaria. The species *Grateloupia filicina* and *G. turuturu* are consumed by Indonesian and Japanese populations, respectively (Zemke-White and Ohno 1999). The alga *Grateloupia turuturu* is also found on the coasts of Europe, where it was introduced, like *Undaria pinnatifida*, with the importation of oyster spat from Japan in the early 1970s. This species has been identified as far north as Portugal (Mulas and Bertocci 2016) and is also found on the French Atlantic coast (see Figure 2.9). However, it is not yet considered invasive on French coasts and therefore does not suffer from a negative image there.

Figure 2.9. Grateloupia turuturu *on the coast of Batz-sur-Mer (Loire-Atlantique, France) (photo credit © Fleurence J. 2023)*

Despite this, *G. turuturu* is not currently authorized for human consumption in France and cannot therefore be marketed as a sea vegetable (Chapter 7, section 7.3.1). The red alga *Chondrus crispus* is also used in human food. As with *Gracilaria* sp., *Chondrus* has two uses, namely as a sea vegetable and as a source of food additives (carrageenans) (Chapter 6, sections 6.2 and 6.7). Other red algae species of the genera *Gelidium* and *Eucheuma* are also exploited as sources of food or phycocolloids (Chapter 6, sections 6.2 and 6.7). Algae belonging to the genus *Palmaria* are used solely as sea vegetables or intermediate food products (Chapter 6, sections 6.2 and 6.3). These algae are mainly found on the northern Atlantic coast, particularly on French, Irish, British, Icelandic, Canadian and American shores. Three species, *Palmaria hecantesis, Palmaria mollis* and *Palmaria palmata*, are consumed as food products. In France, the species *Palmaria palmata* is exploited for this type of application under the name Dulse (Chapter 6, section 6.2).

2.3. Chlorophyceae (green algae)

Unlike brown and red algae, which can be used as food or as sources of phycocolloids for use as food additives (alginates, agar, carrageenans), green algae are mainly used in human nutrition as sea vegetables or intermediate food products (Chapter 6, sections 6.2 and 6.3).

The most commonly consumed green algae belong to the genus *Ulva*. Ulva are ubiquitous algae found on all coasts of the northern and southern hemispheres. These algae are unique in that they are found in coastal areas rich in nutrients, particularly dissolved inorganic nitrogen (NO_3^-, NO_2^-). They are often involved in green tide phenomena, such as *Ulva armoricana* in Brittany (Coat et al. 1998) (see Figure 2.10) and *Ulva prolifera* (Liu et al. 2016) on the coast of Qingdao in China.

Figure 2.10. *Green tide caused by the algae* Ulva armoricana *in Brittany (Hillion Bay, Côtes-d'Armor, France) (photo credit © Fleurence J. 2019)*

Algae harvested for human consumption often belong to other species such as *Ulva lactuca, U. pertusa, U. rigida, U. rotundata* and *U. olivascens*.

Other species of Ulva are valued locally or even regionally; this is particularly the case for *Ulva reticulata,* which is consumed in Vietnam (Zemke-White and Ohno 1999). The species *U. armoricana*, although authorized for human consumption in France like all algae of the genus *Ulva* (Chapter 7, section 7.3), is not currently used for food applications. This is not the case *for Ulva prolifera*, a green tide alga in China that is used for food. This species, also known as *Enteromorpha prolifera* (Algaebase), is consumed in Japan mixed with other green algae such as those belonging to the genus *Monostroma* (Nisizawa et al. 1987). Seven species of *Enteromorpha* are valued as food in Asia. Like Ulva, these algae grow near the shore and are easily accessible at low tide. The most commonly harvested species are *Enteromorpha compressa, E. linza* and *E. intestinalis*. Enteromorphs and Ulva are often used in combination with *Monostroma nitidium* to make Aonori, a food product well known to Japanese consumers. Currently, algae of the genus *Enteromorpha* are included in the genus *Ulva*, and the term *Enteromorpha* is much less commonly used, at least in the current scientific literature.

Green algae belonging to the genus *Codium* (see Figure 2.11) are also consumed in some Asian regions (Table 2.4). Several species of Codiales, such as *Codium bursa, Codium coralloides* and *Codium elongata*, although present on European coasts (Gruet 1989; Cabioc'h et al. 1992), are not used for human consumption.

Figure 2.11. Codium *sp. (photo credit © Fleurence J. 2023)*

Codium species	Consumer country
Codium bartlettii	Philippines
Codium edule	Philippines
Codium fragile	Korea, Philippines
Codium muelleri	Hawaii
Codium taylori	Israel
Codium tenue	Indonesia
Codium tomentosum	Indonesia

Table 2.4. *Examples of Codiales species consumed as food worldwide (Zemke-White and Ohno 1999; Fleurence 2018)*

Among the green algae consumed by humans, it is interesting to note that species grouped in the genus *Caulerpa* are included. For example, the species *Caulerpa taxifolia* (see Figure 2.12), much maligned in Europe for its invasive nature and even toxicity, is traditionally consumed in the Philippines (Zemke-White and Ohno 1999). This alga contains a secondary metabolite, caulerpenyne, which is involved in the chemical defense of *C. taxifolia.* This metabolite appears to be toxic to fauna that graze on it (sea urchins, herbivorous fish), but its production seems to be subject to seasonal variations (Amade and Lemée 1998). This seasonal variation is known to local populations, who only consume the algae at specific times of the year (personal communication, René Perez).

The invasive, non-native nature and presence of caulerpenyne are all obstacles to the development of *Caulerpa taxifolia* as a food source. Several other Caulerpa species are commonly consumed in certain regions of the world, particularly in Asia (Table 2.5).

Species	Country of consumption
Caulerpa taxifolia	Philippines
Caulerpa lentillifera	Philippines
Caulerpa peltata	Philippines
Caulerpa racemosa	Bangladesh, Japan, Philippines, South Pacific islands, Vietnam
Caulerpa sertularioides	Philippines

Table 2.5. *Examples of* Caulerpa *species consumed as food around the world (Zemke-White and Ohno 1999)*

Figure 2.12. Caulerpa taxifolia *(see: https://commons.wikimedia.org/wiki/File:CaulerpaTaxifolia.jpg#/media/File:CaulerpaTaxifolia.jpg)*

2.4. Marine microalgae

Freshwater microalgae (*Chlorella* sp.) and cyanobacteria (*Spirulina* sp.) are microorganisms commonly used in human food as ingredients or food supplements. Marine microalgae are used relatively little compared to spirulina and chlorella. However, three species are used as food supplements. These are the halophilic species *Dunaliella salina*, which lives in salt marshes, and the marine species *Tetraselmis chui* and *Ondotella aurita*.

Dunaliella salina is a microalga belonging to the phylum Chlorophyta (green algae), whose presence is mentioned by Pliny the Elder. He likened it to a salt flower (*Flo salis*) covering the surface of salt marshes (see Figure 2.13). It takes its scientific name from Michel Félix Dunal, professor of botany in Montpellier, who first made scientific observations of this organism in the salt ponds of the Languedoc coast (Barbosa et al. 2023). This alga is particularly well adapted to high variations in salt content (0.5–35%) (Barbosa et al. 2023), showing exceptional tolerance to high salt concentrations. However, this species, which proliferates in salt marshes, can also be found in open marine environments, as is the case in the Black Sea (see Figure 2.13).

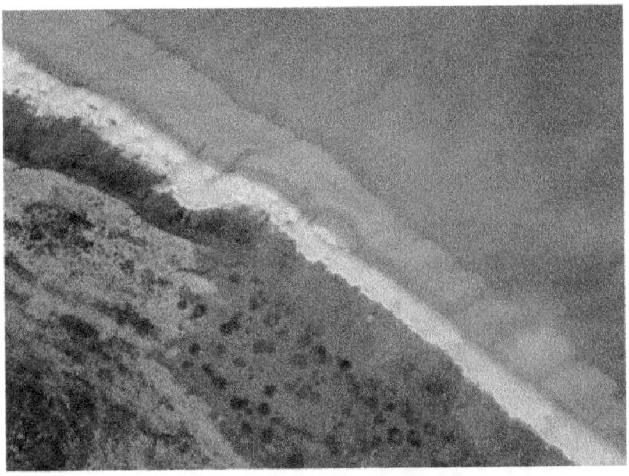

Figure 2.13. *Presence of the microalga* Dunaliella salina *on the shore of Syvach (Black Sea) (see: https://commons.wikimedia.org/w/index.php?curid=107132882)*

Tetraselmis chui is a marine microalga belonging to the phylum Chlorophyta (green algae) and the class Prasinophyceae (Fleurence 2021). It was first isolated in the 1950s on the coast of the United Kingdom and then in various regions of the world. Mainly valued as “feed algae” in aquaculture, it has recently attracted interest for human consumption. Its use as a novel food condiment has been proposed in Canada. The Health and Food Products Directorate there concluded that this microalga has no potential toxicity for food use[1]. In addition to this type of application, this microalga is also marketed as a sea spice. These two types of use are in addition to its use as a dietary supplement.

Odontella aurita is a centric diatom living in the marine environment. It is a brown microalga belonging to the class Bacillariophyceae (Fleurence 2021). This class includes between 11,000 and 100,000 species, depending on the author. *O. aurita* tolerates salinity levels between 27.3 and 35.6 g of $NaCl.L^{-1}$. It is a cosmopolitan species commonly found in the coastal waters of the Atlantic Ocean and the Mediterranean Sea. Its habitat is in the neritic zone, a region corresponding to the continental shelf and preceding the continental slope and abyssal trenches (Box 2.1).

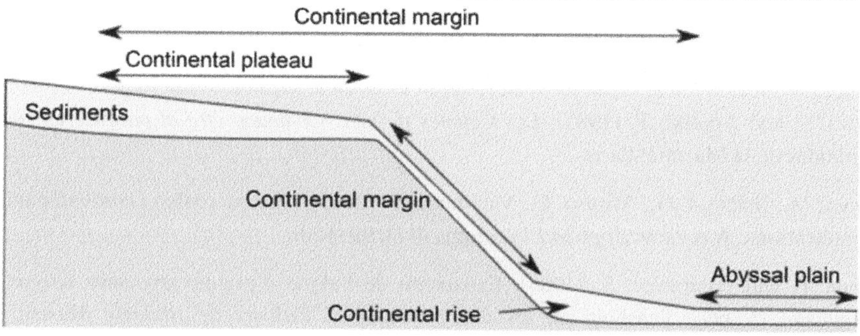

Figure 2.14. *Zoning of the different parts of the ocean (see: https://commons.wikimedia.org/w/index.php?curid=1276487)*

The neritic zone is the part of the ocean that extends from the coastline to the edge of the continental plateau. It is generally characterized by well-oxygenated water and relatively constant temperature and salinity. It also corresponds to the photic zone, the part of the ocean that is penetrated by light and extends to a depth of 200 m below the ocean surface. It is home to a diverse and abundant marine life. As such, it is a prime biotic zone for microphytobenthos and phytoplankton.

Box 2.1. *Distribution of the different coastal and ocean zones*

This microalga, mainly used as a food supplement, is also of interest for the development of new products for human consumption. A patent for the development of plant-based substitutes for meat products includes the addition of *Odontella aurita*

1 See: https://www.canada.ca/fr/sante-canada/services/alimentsnutrition/aliments-genetiquement-modifies-autres-aliments-nouveaux/produits-approuves/renseignements-aliments-nouveaux-tetraselmis-chui-lyophilisee-souche-8-6.html.

(Calleja 2017). This species is also used to make fish egg substitutes using a process protected by a patent application (Calleja and Guillou 2018). *Odontella aurita* is currently one of the most widely used microalgae species for the development of new products outside the usual field of dietary supplements.

Marine microalgae in human food are mainly valued as nutritional supplements, ingredients or substitutes for animal sources. In contrast, marine algae or macroalgae are mainly used as sea vegetables, ingredients or sources of phycocolloids (alginates, agar, carrageenans) (Chapter 6, section 6.7).

2.5. References

Amade, P. and Lemée, R. (1998). Chemical defence of the mediterranean alga *Caulerpa taxifolia*: Variations in caulerpenyne production. *Aquatic Toxycology*, 43, 287–300.

Arasaki, S. and Arasaki, T. (1985). *Les légumes de mer. Comment être et paraître en forme.* Éditions de la Maisnie, Paris.

Barbosa, M., Inácio, L.G., Afonso, C., Maranhão, P. (2023). The microalga *Dunaliella* and its applications: A review. *Applied Phycology*, 4(1), 99–120.

Belsher, T. and Pommelec, S. (1988). Expansion de l'algue d'origine japonaise *Sargassum muticum* (Yendo) Fensholt sur les côtes françaises. *Cahiers de Biologie Marine*, 29, 221–223.

Cabioc'h, J., Floc'h, J.Y., Le Toquin, A., Boudouresque, C.F., Meinez, A., Verlaque, M. (1992). *Guide des algues des mers d'Europe*. Delachaux et Niestlé, Neuchâtel, Paris.

Calleja, P.C. (2017). Substituts végétaux aux produits alimentaires carnés. Patent application national registration number 17 70453.

Calleja, P.C. and Guillou, A. (2018). Substituts d'œufs de poissons. Patent application national registration number 18 70021.

Coat, G., Dion, P., Noailles, M.C., De Reviers, B., Fontaine, J.M., Berger-Perrot, Y., Loiseaux-de Goër, S. (1998). *Ulva armoricana* (Ulvales, Chlorophyta) from the coasts of Brittany (France). II. Nuclear rDNA ITS sequence analysis. *European Journal of Phycology*, 33(1), 81–86.

Deysher, L. and Norton, T.A. (1981). Dispersion and colonization in *Sargassum muticum* (Yendo) Fensholt. *Journal of Experimental Marine Biology Ecology*, 56, 179–195.

FAO (2004). Situation mondiale des pêches et de l'aquaculture [Online]. Available at: https://www.fao.org/4/y5600f/y5600f04.htm.

FAO (2020). Situation mondiale des pêches et de l'aquaculture [Online]. Available at: https://www.fao.org/fsnforum/fr/resources/la-situation-mondiale-des-peches-et-de-laquaculture-2020.

Fleurence, J. (2016). Seaweed as food. In *Seaweed in Health and Disease Prevention*, Fleurence, J. and Levine, I. (eds). Academic Press, Elsevier, Amsterdam.

Fleurence, J. (2018). *Les algues alimentaires : bilan et perspectives*. Lavoisier Tec & Doc, Paris.

Fleurence, J. (2021). *Microalgae: From Future Food to Cellular Factory*. ISTE Ltd, London and John Wiley & Sons, New York.

Fleurence, J. (2024). *Aquatic Biotechnologies: From Genetic Engineering to Enzymatic or Fermentation Engineering*. ISTE Ltd, London and John Wiley & Sons, New York.

Gruet, Y. (1989). Algues. Loire-Atlantique et Vendée. *Bulletin de la Société des Sciences naturelles de l'Ouest de la France-Nantes*.

Guillou, A. and Bossorelle, P. (2022). Composition alimentaire pour la préparation d'un substitut à un produit de la mer. Patent application national registration number 22 07291.

Liu, X., Wang, Z., Zhang, X. (2016). A review of the green tides in the Yellow Sea China. *Marine Environmental Research*, 119, 189–196.

Minchin, D. and Nunn, J. (2014). The invasive brown alga *Undaria pinnatifida* (Harvey) Suringar, 1873 (Laminariales: Alariaceae), spreads northwards in Europe. *Bioinvasions Records*, 3, 57–63.

Mulas, M. and Bertocci, I. (2016). Devil's tongue weed (*Grateloupia turuturu Yamada*) in northen Portugal: Passenger or driver of change in native biodiversity? *Marine Environmental Research*, 118, 1–9.

Murphy, J.T., Johnson, M.P., Viard, F. (2016). A modelling approach to explore the critical environmental parameters influencing the growth and establishment of the invasive seaweed *Undaria pinnatifida* in Europe. *Journal of Theoretical Biology*, 396, 105–115.

Nisizawa, K., Noda, H., Kikuchi, R., Watanabe, T. (1987). The main seaweed foods in Japan. *Hydrobiologia*, 151(152), 5–29.

Ohno, M. and Largo, D.B. (1998). The seaweed resources of Japan. In *Seaweed Resources of the World*, Critchley, T. and Ohno, M (eds). Japan International Cooperation Agency, Tokyo.

Perez, R., Lee, J.Y., Juge, C. (1981). Observations sur la biologie de l'algue japonaise *Undaria pinnatifida* (Harvey) Suringar introduite accidentellement dans l'étang de Thau. *Sciences et Pêches. Bulletin Institut des Pêches Maritimes*, 315, 1–12.

Perez, R., Kaas, R., Barbaroux, O. (1984). Culture expérimentale de l'algue *Undaria pinnatifida* sur les côtes de France. *Sciences et Pêches. Bulletin Institut des Pêches Maritimes*, 343, 3–16.

Veiga, P., Torres, A.C., Rubal, M., Troncoso, J., Sousa-Pinto, I. (2014). The invasive kelp *Undaria pinnatifida* (Laminariales, Ochrophyta) along the north coast of Portugal: Distribution model versus field observations. *Marine Pollution Bulletin*, 84(1/2), 363–365.

Zemke-White, W.L. and Ohno, M. (1999). World seaweed utilisation: An end-of-century summary. *Journal of Applied Phycology*, 11, 69–373.

3

Production Methods

3.1. Context

Global production of algae, both microalgae and macroalgae, has grown steadily over the past 50 years. It rose from 0.56 million tons wet weight in 1950 to 35.82 million tons in 2019 (FAO 2021). However, this growth slowed in the early 2020s (see Figure 3.1) (FAO 2022). The Covid-19 pandemic is likely to be the cause of this slowdown. Most of the world's algae production comes from macroalgae, which accounts for 99% of the total. In 2014, macroalgae production stood at 25 million tons of wet product (FAO 2014). Today, this volume is estimated at nearly 36 million tons, an increase of nearly 40% in just under a decade, despite the slowdown noted in the early 2020s. Only 3% of these volumes come from harvesting in natural fields. In fact, most of the tonnage exploited is obtained through algaculture (Chapter 3, section 3.3).

3.2. Exploitation of natural fields

Global harvesting of macroalgae from natural fields has been declining since the 1990s.

Harvesting in 1990 amounted to 1,330,000 tons of wet weight, compared to 1,080,000 tons in 2019, representing a decrease of approximately 19% in harvested biomass (FAO 2021). Over the past three decades, cultivation has become the main method of producing algal biomass (section 3.3). Natural seaweed beds are harvested manually or mechanically. Manual harvesting is carried out on the foreshore at low tide and is similar to beach fishing (see Figure 3.2).

For a color version of all the figures in this chapter, see www.iste.co.uk/fleurence/marinealgae.zip

This harvesting activity is traditional in Western countries, particularly in France. It involves two types of algae found on the foreshore:

– shore algae or shore seaweed;

– seaweed from shipwrecks or wrack seaweed.

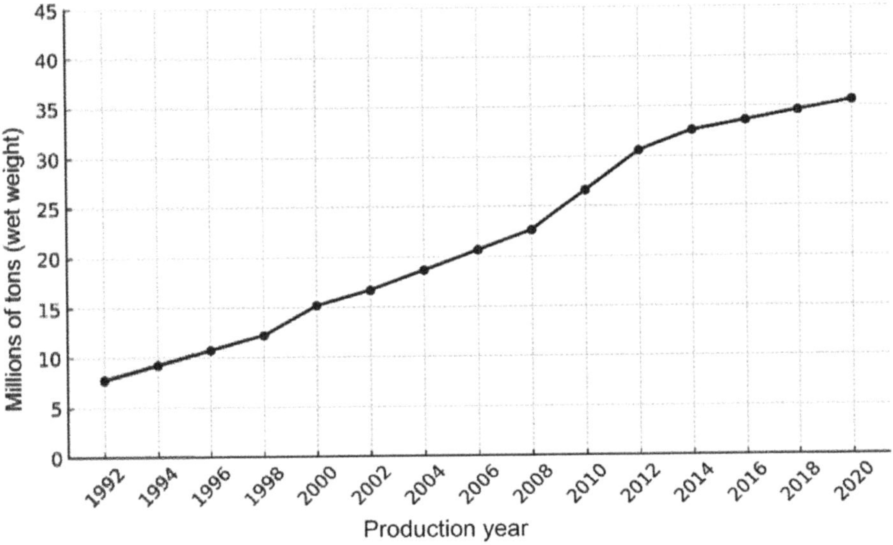

Figure 3.1. *Evolution of global algae production (microalgae and macroalgae) over the last three decades (FAO 2022)*

This distinction has always been important in regulating practices and uses. In Brittany, for example, the "*usement*" of Léon described the rules applicable to the two types of seaweed:

> The Armoriquains of Léon and Dôlas (Daoulas), whose villages and estates border the sea, are each entitled to their land, to enjoy and dispose of the seaweed that is cut from the rocks and other seaweed that the sea washes ashore. However, it is understood that all floating seaweed that is not yet dry belongs to the first person to collect it, whether by boat or by venturing out to the shore to gather it (see: https://fr.wikipedia.org/wiki/R%C3%A9colte_du_go%C3%A9mon).

In order to harmonize seaweed harvesting practices, some of which dated back to feudal times, with different customs depending on the region concerned (Brittany or

Normandy), Colbert issued a royal ordinance on the subject in 1681 (Arzel 1998). This ordinance regulated the harvesting of seaweed (see Article I) and shipwrecks or strandings (see Article V) with the following articles:

> – Article I. The inhabitants of parishes located on the coast shall gather on the first Sunday of January each year, after parish mass, to decide on the days on which the cutting of the seaweed known as *varech*, *goémon*, *vraicq*, *sart* or *gouesmon*, growing in the sea within their territory, shall begin and end.
>
> – Article V. We nevertheless allow any person to take, at any time and in any place, the wrack washed up by the waves on the beaches and to transport it wherever they see fit (Desouches 1972).

Seaweed is harvested from the shores by seaweed gatherers using tools such as seaweed hooks or cutting tools such as guillotines (Arzel 1987).

Figure 3.2. *Harvesting wrack at low tide with a sickle (photo credit © Barbaroux O. 2016)*

This harvesting is carried out at low tide and the seaweed collected, which used to be transported in carts, is now placed in trailers pulled by agricultural vehicles. In France, the species harvested from the shore are mainly brown algae. These include Fucales such as *Fucus vesiculosus* and *F. serratus*. *Fucus vesiculosus* is also known by the common names of "*Varech*" or "*Goémon*" by local populations (Fleurence 2016). Among the frequently harvested shore algae is *Ascophyllum nodosum*, commonly known as *Goémon*. The brown alga *Himanthalia elongata*, well known as *Haricot de mer* (sea spaghetti), is also harvested as a shore alga.

The species *Laminaria digitata* is accessible on foot but only during spring tides (coefficient 100–105); it is harvested at this time to be used as a food seaweed under the name Breton Kombu. However, the same species is harvested using mechanized methods (e.g. scoubidou), mounted on boats, when it is used as a source of food additives such as alginates.

Red algae are also harvested as shore algae. These are mainly species belonging to the genera *Porphyra*, *Chrondrus* and *Gracilaria*. These algae, harvested on European coasts, are mainly intended for human consumption as sea vegetables. Green algae belonging to the genus *Ulva* (*Ulva rotundata, Ulva rigida, Ulva lactuca*) are also harvested as shore algae for food purposes.

The collection of stranded algae has other applications, such as the production of fertilizers, animal feed and the extraction of phycocolloids, such as carrageenans and agar. The harvest of stranded algae of the genus *Gelidium* after winter storms is a source of raw material used in Spain for the supply of agar (Juanes and Sosa 1998). This type of harvesting is also practiced in the French and Spanish Basque Country to obtain a source of carrageenan (*Chondrus crispus*).

In addition to gathering on the foreshore at low tide, there is also a manual seaweed harvesting activity that involves scuba diving with or without specialized equipment (see Figure 3.3). This activity of harvesting seaweed by diving was initially developed in Korea, where women divers harvested seafood and edible seaweed, mainly *Undaria pinnatifida* (Wakame). Since the end of World War II, this activity has spread to many neighboring countries such as Japan, China and the Russian peninsula of Vladivostok (Delaney et al. 2016). In Europe, seaweed is mainly harvested by divers in Portugal and, to a lesser extent, in France, where the red alga *Delesseria sanguinea* is collected for its hemostatic properties (Cabioc'h et al. 1992). In South America, particularly in Chile, agarophytic algae of the genus *Gracilaria*, can also be harvested manually by divers.

Figure 3.3. *Seaweed harvesting by diving in South Korea (photo credit © Barbaroux O. 2016)*

However, manual harvesting of natural fields has given way to mechanized harvesting for certain species of industrial interest. This is particularly the case for the species *Laminaria digitata*, which is used both as a "vegetable" algae and as a source of food additives such as alginates. This species, mainly found on the coast of Brittany, is cut and harvested using a hydraulic claw called a "scoubidou" which is mounted on a vessel known as a "*goémonier*" (seaweed boat) (see Figure 3.4).

Figure 3.4. *Seaweed boat equipped with a "scoubidou" for harvesting* Laminaria *(photo credit © Barbaroux O. 2016)*

In France, the seaweed harvesting fleet has no more than 80 vessels; its size is limited in particular by the number of licenses granted in the Brest maritime district, the main area concerned by Laminaria fields. Nevertheless, a few new or additional licenses were granted (6) between 1985 and 1989 in maritime districts outside Brest (Arzel 1998).

3.3. Algaculture

The main source of algae exploitation is algaculture (97% of global production). The first cultivation trials date back to the 17th century and involved the species *Porphyra tenera* (Perez 1997). Asia is the world's leading producer of algae through algaculture (FAO 2022). Its production accounts for just over 99% of the volumes generated by this activity (see Figure 3.5). Algae farming worldwide also accounts for a significant share of aquaculture activity. It ranks second behind fish farming (see Figure 3.6) (FAO 2022).

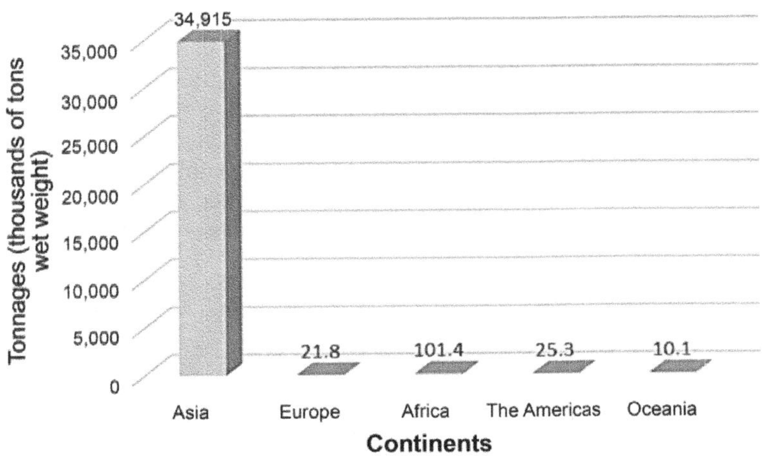

Figure 3.5. *Distribution of global production by algaculture (thousands of tons wet weight) (FAO 2022)*

In 2019, global macroalgae production from aquaculture amounted to approximately 35,000,000 tons (wet weight). In the same year, microalgae production, including spirulina, reached just 56,456 tons of wet weight (FAO 2021). Global algaculture is therefore mainly focused on the production of macroalgae, whether for direct human consumption (sea vegetables, ingredients) or indirect consumption (phycocolloids). The distribution of this production differs according to botanical groups (brown algae, red algae, green algae) (see Figure 3.7).

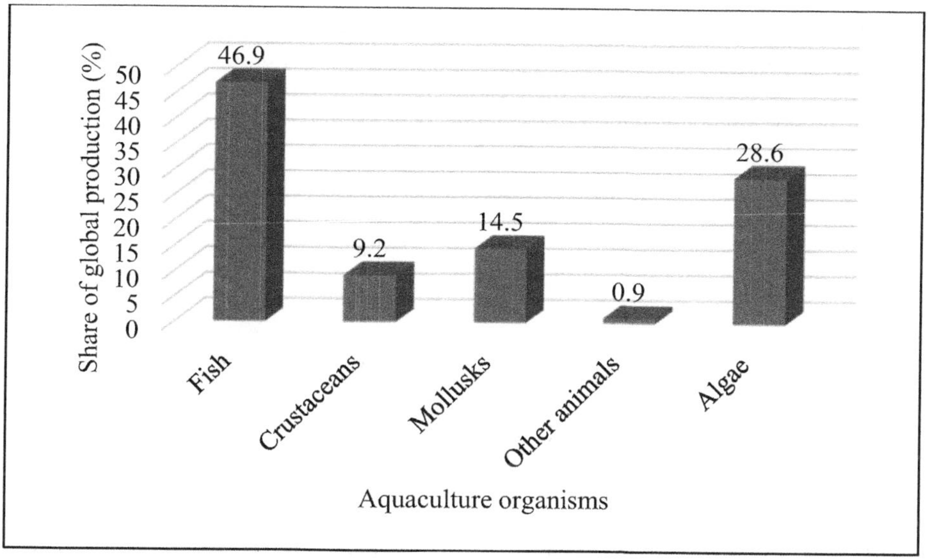

Figure 3.6. *Global share of seaweed production by aquaculture in 2020 compared to other production (fish, crustaceans, mollusks, other animals) (FAO 2022)*

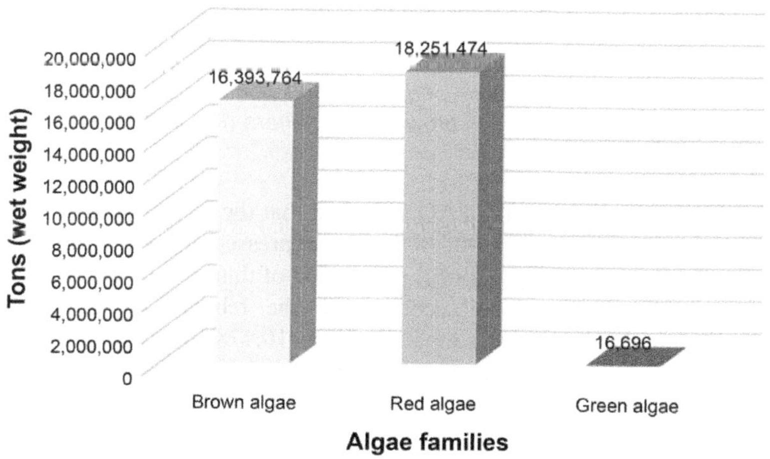

Figure 3.7. *Distribution in 2019 of global aquaculture production of different botanical groups of algae (brown, red or green algae) (FAO 2021)*

The most widely produced species belong to the genus *Laminaria*. Among these, the main tonnages concern the species *Laminaria japonica*, now known as *Saccharina japonica*. Algae of the genus *Laminaria/Saccharina* represent nearly 75% of brown algae aquaculture production (see Figure 3.8). They are used as sea vegetables under the generic name Kombu or as a source of alginates for the food industry (Chapter 6, sections 6.2 and 6.7).

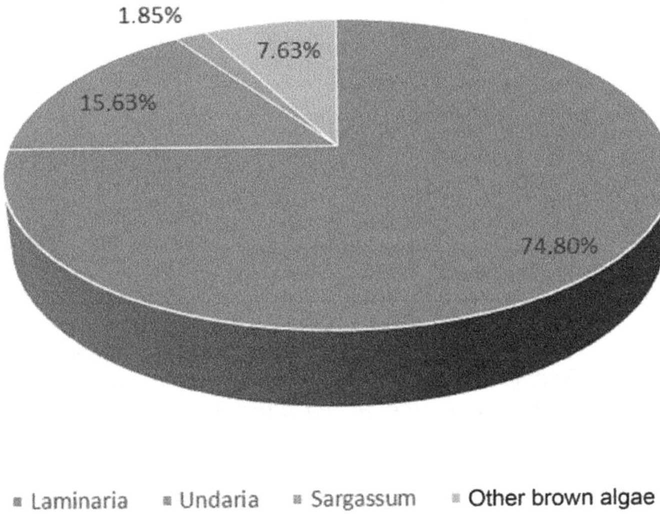

Figure 3.8. *Distribution of global production by algaculture in 2019 of the main brown algae genera (FAO 2021)*

Data published in 2014 by the FAO indicated that the algacultural production of *Laminaria japonica* stood at 6,900,000 tons expressed in wet weight. Figures published in 2021 for 2019 show global production of this species at 12,273,748 wet tons. *Laminaria japonica* alone accounts for the rebound in global seaweed production through aquaculture. China produces 10,978,362 tons, accounting for nearly 90% of global production of *L. japonica* (FAO 2021). *Undaria pinnatifida*, or Wakame, is the second most commonly cultivated brown algae species. It accounts for 15.3% of global production of brown algae for food. As with *Laminaria japonica,* China is the main producer of this species through algaculture (see Figure 3.9). It is also the leading producer of red algae of the genus *Porphyra*, recently renamed *Pyropia*. This red alga is frequently consumed in Asia and now worldwide under the common name Nori (Chapter 6, section 6.3).

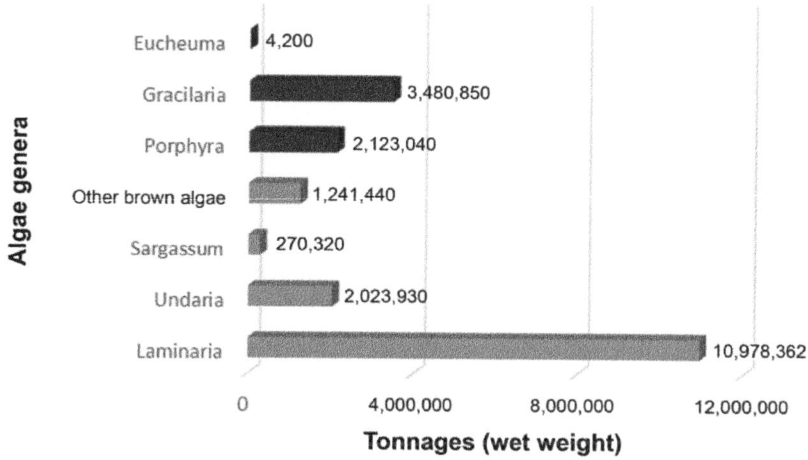

Figure 3.9. *Distribution of Chinese production in 2019 by alga genus (FAO 2021)*

In addition to *Porphyra*, which is mainly used as a sea vegetable, other red algae are cultivated for the production of phycocolloids, particularly those of the carrageenan family. Global production of carrageenophytes through algaculture is estimated at 11,622,213 tons (wet weight). China contributes very little to this production (4,200 tons). Indonesia remains the main producer, but the figures available combine cultivated and harvested resources. The world's leading producer of carrageenophytes is the Philippines.

The species cultivated in the archipelago are *Kappaphycus alvarezii* and *Eucheuma denticulatum*. They are sources of κ- and ι-carrageenans, respectively (see Chapter 6, section 6.7) (Trono 1998). The tonnages produced are 1,423,168 tons for *K. alvarezii* and 75,619 tons for *E. denticulatum* (FAO 2021).

Agar-producing red algae are also cultivated. Global production of these algae is estimated at nearly 3,700,000 tons (wet weight) (FAO 2021). China, with a production of 3,480,850 tons, is the leading producer of agarophytes through cultivation.

Green algae are rarely cultivated. The FAO estimated global production of Chlorophyceae at 16,696 tons (wet weight) in 2019 (FAO 2021). This production is associated with three groups, one of which is not precisely defined (see Figure 3.10).

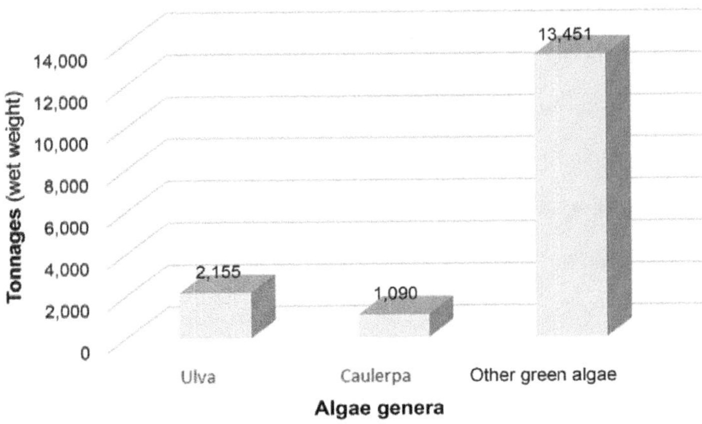

Figure 3.10. *Distribution of global production by the type of green algae (FAO 2021)*

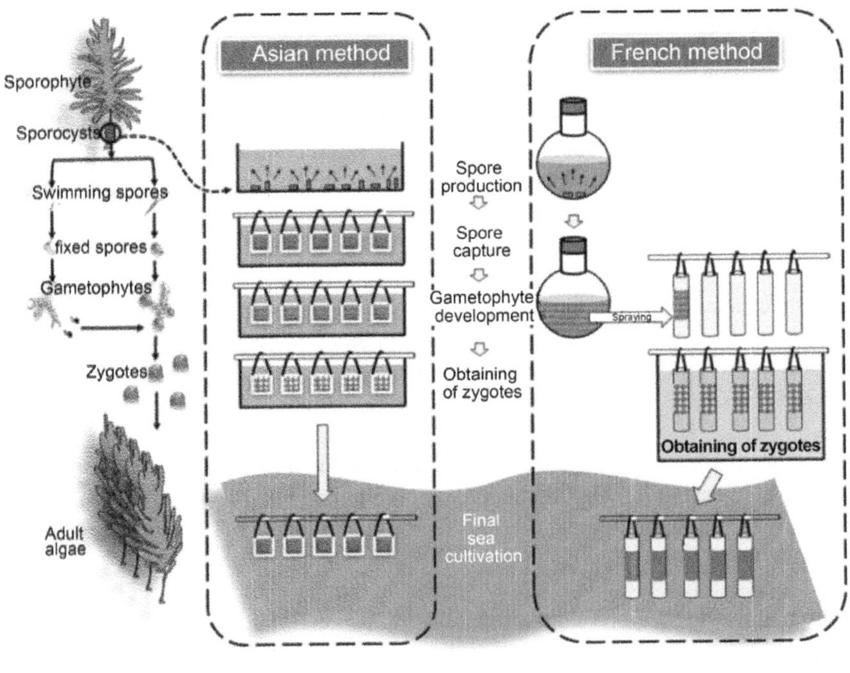

Figure 3.11. *Comparison of the Asian method and the French method for cultivating the brown alga* Undaria pinnatifida *(Wakame) (Fleurence 2018)*

Algae cultivation relies on knowledge and mastery of the species' biological cycle. The life cycle of brown algae such as *Laminaria japonica* or *Undaria pinnatifida* has been particularly well known for a very long time. The species *Undaria pinnatifida*, valued as a form of food algae in both Asia and Europe, are cultivated on both continents.

In France, *Undaria* cultivation was developed in the mid-1990s under the impetus of the Ifremer algology laboratory. The cultivation technique developed, now known as the "French method", differs significantly from the traditional method practiced in Asia (see Figure 3.11).

Unlike the Asian method, the French method involves fewer steps and the production of spores, followed by gametophytes, in a limited volume environment that is better controlled, particularly in terms of temperature (see Figure 3.11). This method, known as free-living cultivation, is based on a number of initial steps carried out in the laboratory (see Figure 3.12).

The first of these steps involves selecting a few dark fragments of algae from the sporiferous zone. These fragments are the most prolific in spores. After cleaning these fertile base fragments, they are immersed in a container of seawater in which the sporocysts burst, releasing swimming spores at a density of 30,000–40,000 per mL (Perez 1997).

This step, which takes place at 15°C in a limited volume, usually a laboratory flask, produces a concentrated spore preparation. After filtering the solution, the filtrate is poured into a flask containing a nutrient medium rich in salts, which causes a whitish precipitate to form. The presence of this precipitate is essential, as the spores need to attach themselves in order to germinate. It is during attachment to the precipitate that the spore germinates into a male or female gametophyte.

After several days of incubation, the precipitate is dissolved by injecting carbon dioxide and the gametophytes are released into the medium, hence the term "freeliving", which was given to the French method.

The temperature is then raised in increments of 0.5°C up to 21°C, producing growth and multiplication of the gametophytes without inducing gamete release. The medium darkens under the density of the gametophytes produced. This balloon can be used as a reserve for the production of seed balloons awaiting use or for seeding.

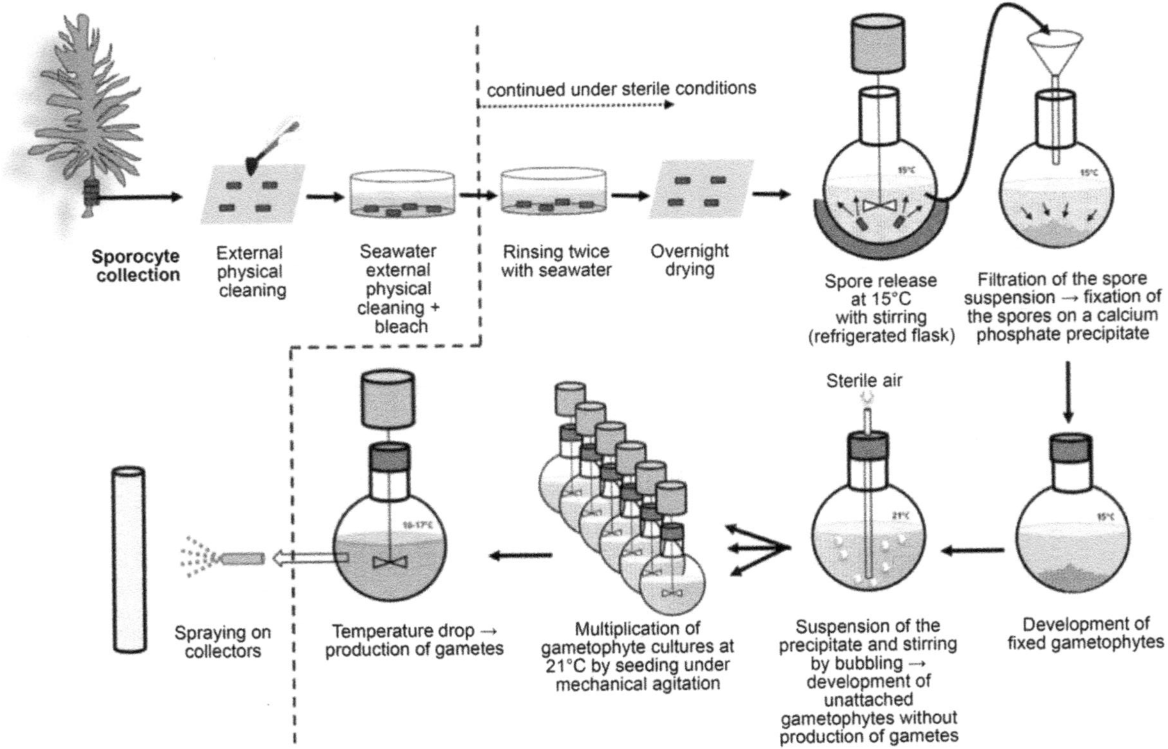

Figure 3.12. *Production of gametophytes and gametes of* Undaria pinnatifida *using the "free-living" method (Fleurence 2018)*

The final stage, or seeding, begins with a drop in temperature from 21 to 16 or 17°C. This drop in temperature triggers the fertility of the gametophytes, which release 10^{12} gametes, whose probability of fusion to form a zygote is optimized by the confined environment of the 6 L balloon. The fertile solution is then sprayed directly onto collectors in development tanks and, three weeks later, the strings bear 5 mm seedlings, which are then placed in the sea.

Compared to the Asian method, the French method of free living means stocks of gametophytes can be created that are available all year round, as they are stored in balloons kept in the laboratory. It means seeding can take place at any time of year on the sites, particularly in the event of destruction by storms. It also requires less labor than the Asian method, particularly during the first stage of cultivation.

The Asian method involves collecting large quantities of fertile bases, which requires a lot of labor. These fertile bases are then immersed in cement basins, where the sporocysts are broken open by agitating the water with the help of numerous people dedicated to this task (Perez 1997).

Microalgae cultivated for food purposes mainly belong to the genus *Spirulina*, currently renamed *Arthrospira* (Spirulina). Although Spirulina is a cyanobacterium, it is still considered by many in the industry to be a microalga. As such, it is included in the FAO's global statistics under the heading "microalgae production".

Spirulina is a cyanobacterium that can be cultivated in fresh or salt water. It is generally cultivated in fresh water, but information on production conditions (fresh water, salt water) is not always available in international statistics. Of the approximately 56,500 tons of microalgae produced worldwide in 2019, Spirulina cultivation accounted for 99% of this production (FAO 2021). Just under 1%, or 248 tons in wet weight, was associated with the production of green microalgae (*Haematococcus* sp., *Chlorella* sp., *Odontella* sp., *Dunaliella* sp.).

As with macroalgae, China is the leading producer of microalgae, particularly Spirulina (see Figure 3.13).

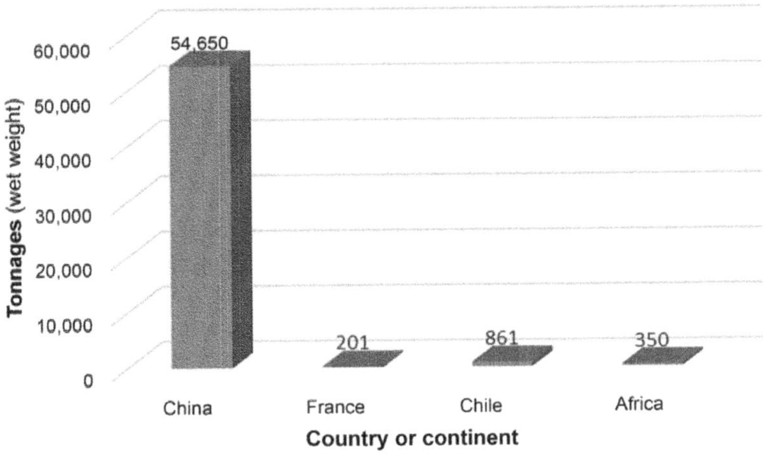

Figure 3.13. *Main producing countries of Spirulina (Spirulina/*Arthrospira*) (FAO 2021)*

In Europe, France is the leading producer of Spirulina, accounting for 58% of European production (see Figure 3.14) (FAO 2021).

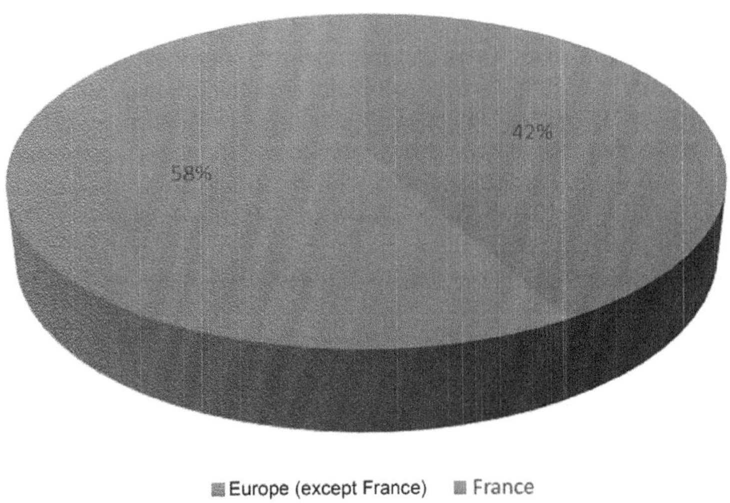

Figure 3.14. *France's share of European production by Spirulina cultivation (FAO 2021)*

France also remains the European leader in the production of microalgae other than Spirulina. In fact, the 6 tons of green microalgae (*Chlorella* spp.) cultivated in Europe in 2019 originated in France. French aquaculture production is divided between the genera *Spirulina*, *Chlorella* and *Tetraselmis*, with very different production levels (Table 3.1).

Genus or species	Tons (wet weight)
Spirulina spp. (Cyanobacteria)	201
Chlorella vulgaris (green algae)	5
Tetraselmis spp. (brown diatom algae)	1

Table 3.1. *Distribution of French microalgae production by genus or species (FAO 2021)*

The African continent is also a producer of Spirulina, although its volume is limited compared to Chinese production, but higher than European production (see Figure 3.13). Chile, with a production of 861 tons of Spirulina, is the leader in the cultivation of this organism on the South American continent.

Microalgae can be cultivated in open environments or closed circuits (photobioreactors). For food applications, cultivation mainly takes place in open environments in natural (ponds) or artificial basins. The artificial basins used are often raceway type basins (see Figure 3.15). They have a distinctive geometric shape similar to that of a racetrack or hippodrome, hence the term "raceway". Water circulation, and therefore culture circulation, is ensured by a paddle wheel. In this culture method, nutrients and water are supplied continuously and the biomass produced is harvested as it develops. The purpose of this removal is to prevent the upper microalgal layer from shading the lower layers that are still developing. Raceway ponds can be open to the air or protected by blackout curtains inside greenhouses (see Figure 3.16).

Raceway cultivation is widely used for the production of *Spirulina* sp., *Dunaliella* sp. and *Haematococcus* sp. (Fleurence 2021).

Among the microalgae living in salt water and with potential for use in human food is the alga *Porphyridium cruentum* (see Figure 3.17).

Figure 3.15. *Raceway tank used for* Odontella aurita *microalgae production (photo credit © Braud J.-P. 2020)*

Figure 3.16. *Raceway-type tank for cultivating Spirulina inside a greenhouse and protected by blackout curtains to control incident light (photo credit © Westspiruline, Velly S. 2019)*

This unicellular red alga belongs to the same botanical family (Bangiales) as *Porphyra* sp. (Nori), which is already used in human food. Like *Porphyra*, this microalga is a significant source of protein. Its content can reach up to 56% of the algal biomass (Safi et al. 2013). It also contains a protein pigment, B-phycoerythrin (BPE), which is easily extractable and potentially valuable as a natural colorant (Jubeau et al. 2013). This alga is mainly cultivated in closed systems in tubular photobioreactors for high value-added applications (pigments, biodiesel) (Fleurence 2021). Historically, its cultivation was developed in France by the CEA to produce this microalga for food and B-phycoerythrin valorization (see Figure 3.18).

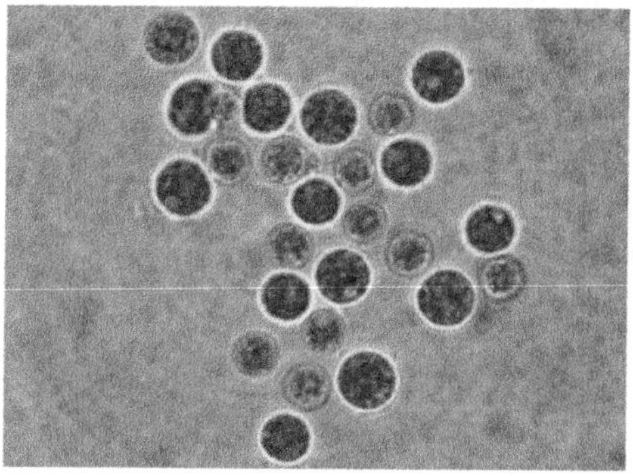

Figure 3.17. *Unicellular alga* Porphyridium cruentum *(photo credit © Jubeau S. 2020)*

Figure 3.18. *Cultivation of the microalga* Porphyridium cruentum *in tubular photobioreactors set up at CEA Cadarache (photo credit © Braud J.-P. 1990)*

Unfortunately, the application for authorization to use this microalga in human food was rejected by the French High Council for Public Health. This opinion, mainly motivated by the lack of previous consumption of *P. cruentum* by human populations, put an end to the development of this alga as a food source in France (personal communication, Joël Fleurence 1992).

Despite this, *Porphyridium cruentum* remains a potentially interesting microalga for food use, along with Spirulina, as a source of protein, pigments, polysaccharides and fatty acids (Fleurence 2021). Its cultivation is well established, both technically and economically. Only the regulatory aspect, in France but not in Europe, remains to be redefined.

Algae are mainly harvested worldwide through cultivation and are used directly or indirectly (as colloids) in food. The biomass produced is predominantly macroalgae, with microalgae being produced mainly to meet the relatively limited demand for food supplements (Chapter 6, section 6.4).

3.4. References

Arzel, P. (1987). *Les goémoniers*. Le chasse-marée – Éditions de l'estran, Douarnenez.

Arzel, P. (1998). *Les laminaires sur les côtes bretonnes : évolution de l'exploitation et de la flottille de pêche, état actuel et perspectives*. Ifremer, Plouzané.

Cabioc'h, J., Floc'h, J.-Y., Le Toquin, A., Boudouresque, F.-C., Meinesz, A., Verlaque, M. (1992). *Guide des algues des mers d'Europe*. Delachaux et Niestlé, Neuchâtel, Paris.

Delaney, A., Frangoudes, K., Li, S.A. (2016). Society and seaweed: Understanding the past and present. In *Seaweed in Health and Disease Prevention*, Fleurence, J. and Levine, I. (eds). Academic Press, Elsevier, Amsterdam.

Desouches, M.J. (1972). La récolte de goémon et l'ordonnance de la Marine. *Annales de Bretagne et des pays de l'Ouest* 79(2), 349–371.

FAO (2014). La Situation mondiale des pêches et de l'aquaculture en 2014 [Online]. Available at: https://openknowledge.fao.org/items/c336fbd6-b2c5-40db-87b6-d0545eb15515.

FAO (2021). Global seaweeds and microalgae production, 1950-2019 [Online]. Available at: https://openknowledge.fao.org/server/api/core/bitstreams/97409d09-2f8e-4712-b11e-60105d89959b/content.

FAO (2022). La Situation mondiale des pêches et de l'aquaculture en 2022 [Online]. Available at: https://openknowledge.fao.org/server/api/core/bitstreams/00f1704f-b092-492c-84da-63c2a88923e2/content/cc0461fr.html.

Fleurence, J. (2016). Seaweed as food. In *Seaweed in Health and Disease Prevention*, Fleurence, J. and Levine, I. (eds). Academic Press, Elsevier, Amsterdam.

Fleurence, J. (2018). *Les algues en alimentation humaine. Bilan et perspectives*. Lavoisier, Paris.

Fleurence, J. (2021). *Microalgae: From Future Food to Cellular Factory*. ISTE Ltd, London and John Wiley & Sons, New York.

Juanes, J.A. and Sosa, P.A. (1998). The seaweed resources of Spain. In *Seaweed Resources of the World*, Critchley, T. and Ohno, M. (eds). Japan International Cooperation Agency, Tokyo.

Jubeau, S., Marchal, L., Pruvost, J., Jaouen, P., Legrand, J., Fleurence, J. (2013). High pressure disruption: A two-step treatment for selective extraction of intracellular components from the microalga *Porphyridium cruentum*. *Journal of Applied Phycology*, 25, 983–989.

Perez, R. (1997). *Ces algues qui nous entourent*. Ifremer, Plouzané.

Safi, C., Charton, M., Pignolet, O., Pontalier, P.Y., Vaca-Garcia, C. (2013). Evaluation of the protein quality of *Porphyridium cruentum*. *Journal of Applied Phycology*, 25, 497–501.

Trono, Jr G.C. (1998). The seaweed resources of the Philippines. In *Seaweed Resources of the World*, Critchley, T. and Ohno, M. (eds). Japan International Cooperation Agency, Tokyo.

4

Biochemical Composition and Nutritional Value of Marine Algae

4.1. Introduction

The biochemical composition of marine algae varies depending on the species and differs according to the season for a given species (Morgan et al. 1980; Denis et al. 2010). This high variability affects carbohydrate levels, which are mainly present in the form of polysaccharides that behave like dietary fiber (Lahaye 1991). It also affects protein content, which varies greatly depending on the botanical family concerned. Red algae have the highest protein content, namely 47% of dry matter for the species *Porphyra tenera* (Fujiwara-Arasaki et al. 1984; Dumay and Morançais 2016), and are also greatly affected by seasonal variation in their protein fraction.

The lipid fraction of marine algae, although representing a limited proportion of dry matter (1–6%) (Wielgosz-Collin et al. 2016), is also subject to inter-species and annual variation (Fleurence et al. 1994; Kendel et al. 2013). Apart from interspecificity and season, other abiotic factors (light intensity, photoperiod, water temperature) play a regulatory role in the production of certain proteins such as R-phycoerythrin in red algae and therefore have an impact on the protein fraction of these algae. The carbohydrate, protein and lipid contents of algae and their bioavailability help determine their nutritional value and the importance of this marine resource for human and animal consumption.

For a color version of all the figures in this chapter, see www.iste.co.uk/fleurence/marinealgae.zip

Marine Algae, A Sustainable Food Resource,
by Joël FLEURENCE.

4.2. Carbohydrates

In algae, carbohydrates are mainly present in the form of polymers or polysaccharides. These polysaccharides have a structural or storage function. Three types of polysaccharides are traditionally classified. From a structural point of view, a distinction is made between polysaccharides that form the skeleton of the cell wall, or fibrillar polysaccharides, and polysaccharides that make up the parietal matrix, or matrix polysaccharides. In addition to these two types of carbohydrate polymers, there is also a category of storage polysaccharides (Table 4.1) (Arasaki and Arasaki 1985; Stiger-Pouvreau et al. 2016). The biochemical nature of these carbohydrates differs depending on the species. The carbohydrate content also differs between species and varies within a given species depending on the season and algaes' stage of development.

Polysaccharides	Brown algae (Pheophyceae)	Red algae (Rhodophyceae)	Green algae (Chlorophyceae)
Parietal skeleton (skeletal phase)	Cellulose Hemicellulose	Cellulose Hemicellulose	Cellulose Hemicellulose
Parietal matrix (matrix phase or mucilage)	Alginates Fucans	Sulfated galactans: – carrageenans – agar – porphyran	Neutral polysaccharides Sulfated polysaccharides (ulvanes)
Storage	Laminarin	Floridene starch	Starch Fructan

Table 4.1. *Examples of the distribution of algal polysaccharides and their nature, according to algal botanical families (Arasaki and Arasaki 1985; Stiger-Pouvreau et al. 2016)*

4.2.1. *Carbohydrates in brown algae (Phaeophyceae)*

The carbohydrate content of brown algae varies between 38 and 68% of dry matter (Darcy-Vrillon 1993). Brown algae such as Laminariales (*Laminaria digitata*) or Fucales (*Fucus vesiculosus*) contain cellulose as a structural polysaccharide of the cell wall (Table 4.1). This polymer can represent up to 20% of the dry matter of the algal thallus (Kloareg and Quatrano 1988). However, this cellulose is rarely a homogeneous polymer of β-1-4-glucose, but may contain other sugars such as xylose in its molecular structure (Kloareg and Quatrano 1988).

4.2.1.1. *Alginic acid (alginates)*

Laminariales (*Laminaria digitata* and *Saccharina japonica*) and Fucales (*Fucus serratus*, *Ascophyllum nodosum*) are also characterized by the predominant presence of alginic acid in the matrix or mucilage phase. This acid is best known as "alginate", which is the ionized form of alginic acid. Alginates account for between 10 and 45% of the dry matter of these algae (Kloareg and Quatrano 1988). The basic structure of alginates is a polymer formed by the combination of mannuronic acid monomers and guluronic acid monomers linked together by 1,4-O-glycosidic bonds. These monomers form homogeneous blocks of mannuronic or guluronic acid (see Figure 4.1) called Poly M or Poly G. Alginic acid is a heteropolymer resulting from the alternating chaining of Poly M and Poly G blocks. This combination leads to the formation of a long chain consisting of 1,500–2,000 monomeric units with an average molecular weight of 800 kDa (Perez 1997).

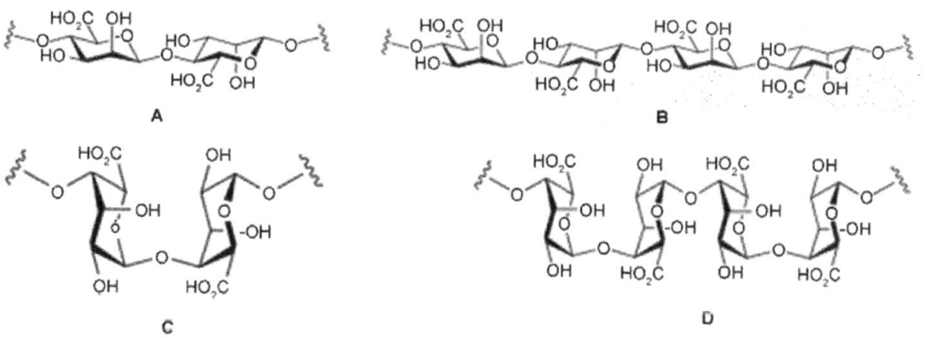

Figure 4.1. *Structure of mannuronic acid (A) and guluronic acid (C) blocks that make up the alginic acid polymer [Poly M (B)-Poly G (D)] (Fleurence 2018)*

α-1,4-L-guluronic acid is produced from β-1,4-D-mannuronic acid via an epimerization reaction catalyzed by an enzyme, C5 epimerase. The ratio of mannuronic acid to guluronic acid (MA/GA) can vary from 0.25 to 2.25 depending on the species, the parts of the thallus and the age of the algae (Kloareg and Quatrano 1988) (Table 4.2). Thus, in the species *Laminaria digitata*, exploited in France for the production of alginates, the MA/GA ratio of young fronds reaches a value of 2.3 compared to 1.35 for senescent fronds (Haug et al. 1974). This difference is even more pronounced in the species *Fucus vesiculosus*, where the MA/GA ratio is 2.2 for young tissues and only 0.6 for aged tissues (Haug et al. 1974).

The MA/GA ratio is indicative of the rheological properties, particularly the gelling properties, of alginates (Perez 1997) (Chapter 6, section 6.7).

Species	MA/GA ratio
Pelvetia caniculata (Fucale)	1.5
Fucus serratus (Fucale)	1.3
Ascophyllum nodosum (Fucale)	1.8
Laminaria digitata (Laminaria)	1.4
Laminaria hyperborea (Laminaria)	1.35
Laminaria diabolica (Laminaria)	1.12
Undaria pinnatifida (Laminariaceae)	0.75
Desmarestia aculeata (Desmarestiale)	0.85
Sargassum sagamianum (Fucale)	0.44
Hizikia fusiformis (Fucale)	0.85

Table 4.2. *Distribution of mannuronic and guluronic acids (MA/GA ratios) according to different species of brown algae (Haug et al. 1974; Mizuno et al. 1983)*

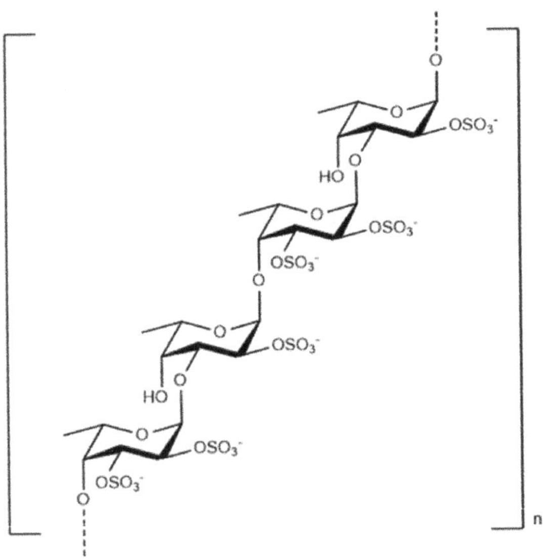

Figure 4.2. *Example of the chemical structure of fucoidan from* Ascophyllum nodosum *(Ale and Meyer 2013; Fleurence 2018)*

4.2.1.2. *Fucans*

Brown algae also contain fucans as matrix polysaccharides. These are polymers mainly composed of sulfated L-fucose (Kloareg and Quatrano 1988). However, this family of polymers is very heterogeneous, and at least three types of fucans can be distinguished: fucoidans (see Figures 4.2 and 4.3), ascophyllans and glucuronogalactofucans or glucuronofucogalactans. These three types of fucans may or may not coexist in the thallus of algae, but their relative proportions differ depending on the species (Table 4.3).

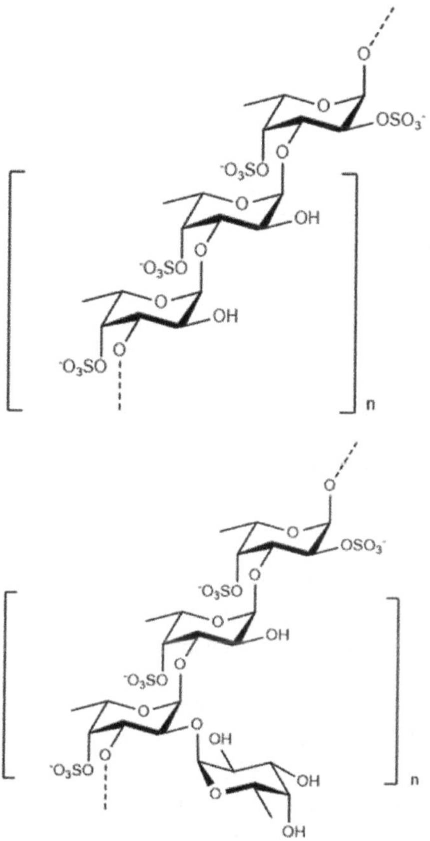

Figure 4.3. *Examples of the chemical structure of fucoidan from* Saccharina latissima (Laminaria saccharina) *and* Ecklonia *sp. (Ale and Meyer 2013; Fleurence 2018)*

Fucoidans are a group of heterogeneous molecules that can be distinguished, in particular, by the position of the sulfate groups on either side of the α 1-3 L-fucopyranose skeleton (see Figures 4.2 and 4.3).

Algae genus	Fucoidans	Ascophyllans	Glucuronofucogalactans
Laminaria sp. (Laminariale)	++		
Alaria sp. (Laminariale)		+	+
Macrocystis sp. (Laminariale)	++		
Undaria sp. (Laminariale)		+	+
Eisenia sp. (Laminariale)	++	+	
Ascophyllum sp. (Fucale)	+	++	+
Fucus sp. (Fucale)	+	+	
Himanthalia sp. (Fucale)	+	+	
Hizikia sp. (Fucale)		++	
Sargassum sp. (Fucale)		+	++

Table 4.3. *Distribution of different types of fucans according to brown algae genera and families (Kloareg and Quatrano 1988)*

The presence of unsulfated fucose residues or glucuronic acids on the basic structure of fucoidans is also observed in *Laminaria saccharina* (*Saccharina latissima*) and *Cladosiphon okamuranus* (Ale and Meyer 2013).

Ascophyllan is a fucan initially isolated from the brown alga *Ascophyllum nodosum* (Medcalf and Larsen 1977a). The basic structure of ascophyllan consists of a main skeleton resulting from the linking of glucuronic, guluronic and mannuronic acids. Side chains containing fucose, xylose and sulfated groups are grafted onto this basic structure to form the final structure (Larsen et al. 1966, 1970).

Ascophyllans can also be associated with proteins and form an ascophyllan-like complex, in which the proteins act as molecular bridges between the ascophyllans and other matrix fucans (Medcalf and Larsen 1977b).

4.2.1.3. *Laminarin*

In addition to polysaccharides with a structural function, brown algae also contain reserve or storage polysaccharides. Laminarin fulfils this function in brown algae that are frequently used in human food. This storage polysaccharide was first identified in Laminaria, hence the name laminarin. In species of the genus *Laminaria,* laminarin can account for 2–34% of the algaes' dry matter (Stiger-Pouvreau et al. 2016). It is a linear polysaccharide consisting of 25–50 glucose monomers linked together by β-1-3 glycosidic bonds (see Figure 4.4). Two types of laminarin are distinguished: those ending with mannitol at the end of the chain, known as laminarin M, and those ending with glucose, known as laminarin G (see Figure 4.5) (Percival 1979). Laminarin has an average molecular weight of 5 kDa (Stiger-Pouvreau et al. 2016).

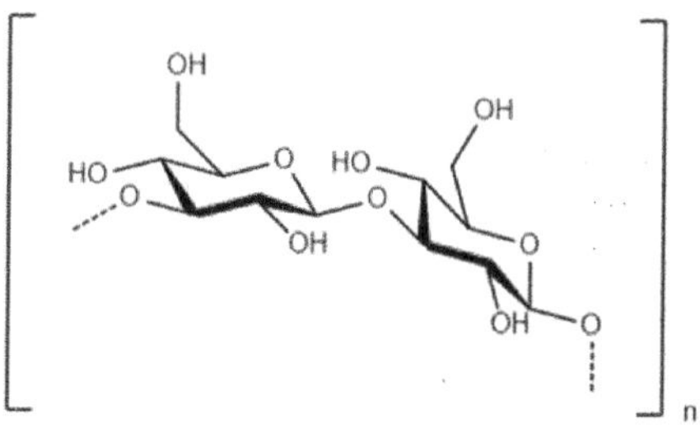

Figure 4.4. *Structure of laminarin (Percival 1979; Fleurence 2018)*

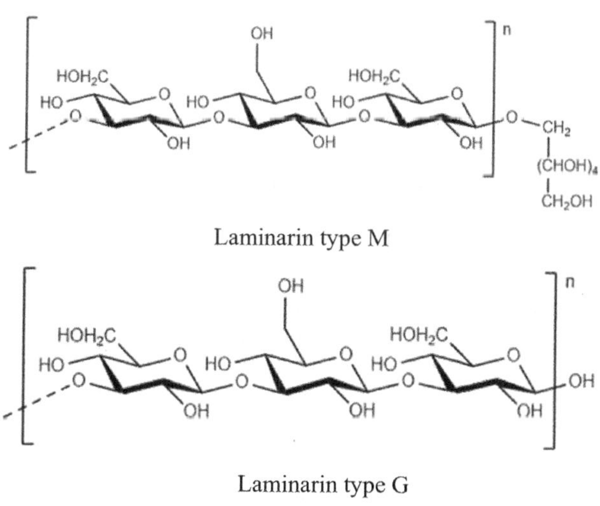

Figure 4.5. *Structure of laminarins M and G (Percival 1979; Fleurence 2018)*

4.2.2. *Carbohydrates in red algae (Rhodophyceae)*

As with Phaeophyceae, the carbohydrate content of red algae varies depending on the species and the season for a given species (Table 4.4).

Species	Carbohydrate content (% of dry matter)
Palmaria palmata	38–74
Gracilaria verrucosa	55–61
Chondrus crispus	50–60
Porphyra tenera	44–55

Table 4.4. *Examples of variations in the carbohydrate content of some red algae valued as sea vegetables (Morgan et al. 1980; Darcy-Vrillon 1993)*

Carbohydrates in red algae are mainly in polymerized or polysaccharide form. As in brown algae, there are fibrillar polysaccharides that form the parietal skeleton, matrix polysaccharides that make up the parietal matrix and storage polysaccharides.

In red algae, parietal fibrillar polysaccharides are mainly cellulose, mannan polymers or xylans (Deniaud 2002; Stiger-Pouvreau et al. 2016). Mannans result from the linear linkage of β-D-(1,4) mannose, and xylans are linear polymers of β-D-(1,3) xylose or β-D-(1,4) xylose (Deniaud 2002). In some red algae, the presence of hemicellulose in small quantities (4% of dry matter) has also been reported in addition to the main fibrillar polysaccharides mentioned above.

In the species *Kappaphycus alvarezii*, which is used for the production of carrageenan-type food additives (Chapter 6, section 6.7), the presence of a sulfated glucan and a glucomannan has been described (Lechat et al. 2000).

The matrix polysaccharides of red algae are galactans with varying degrees of sulfation, the nature of which varies depending on the species (Deniaud 2002). Red algae also contain a storage polysaccharide or floridene starch (Stiger-Pouvreau et al. 2016), better known by its common name rhodamylon.

The three main families of matrix polysaccharides in red algae are agar, carrageenan and porphyran. These polysaccharides are present in specific and predominant quantities in certain species known as agarophytes, carrageenophytes or porphyranophytes.

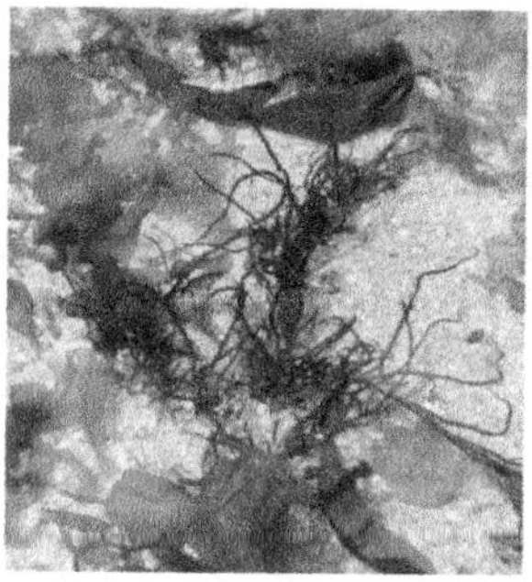

Figure 4.6. Gracilaria verrucosa, *agarophyte red alga (photo credit © Ragueneau E. 2024)*

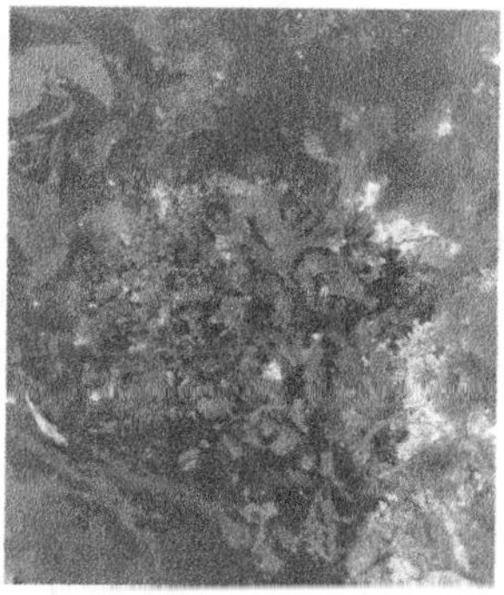

Figure 4.7. Chondrus crispus *(Irish moss), carrageenophyte red alga (photo credit © Ragueneau E. 2024)*

Figure 4.8. Porphyra umbilicalis, *porphyranophyte red alga (photo credit © Fleurence J. 2023, personal algae collection)*

Box 4.1. *Rhodophyceae (red algae), various sources of original matrix polysaccharides*

4.2.2.1. *Agar (agar-agar)*

Red algae of the genus *Gelidium* or *Gracilaria* are known as agarophytes. They contain a mucilaginous polymer called "agar", better known as agar-agar. Agar-agar is a complex composed of two main elements, agarose and agaropectin, in a ratio of 1:3 (Arasaki and Arasaki 1985).

Agarose is a neutral polymer with a linear structure consisting of repeated chains of a disaccharide, itself composed of a molecule of β-D-galactose and 3,6-α-L-anhydrogalactose (see Figure 4.9).

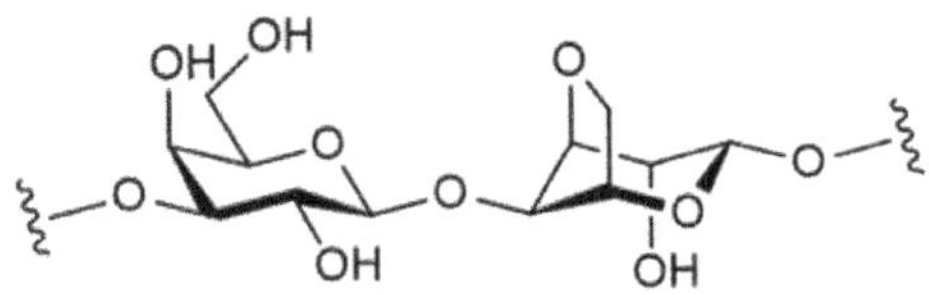

β-D-galactose   3,6-α-L-anhydrogalactose

Figure 4.9. *Structure of the agarose disaccharide*

Agaropectin is an acid-polysaccharide that differs from agarose by the presence of sulfated ether groups at C6 of D-galactose (see Figure 4.10). The degree of sulfation can reach 20% of the molecular composition. However, the sulfate content of the final agar-agar polymer (agarose + agaropectin) remains fairly low, ranging between 3.9 and 9.7% of the total elemental composition.

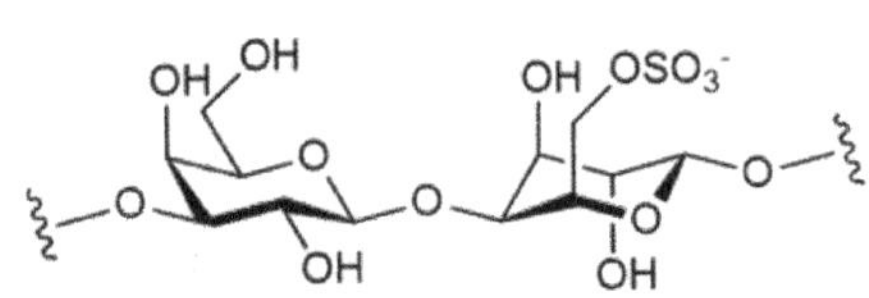

Figure 4.10. *Structure of agaropectin*

Other substituents such as pyruvate groups (3%), methyl groups or sometimes glucuronic acid residues may also be present in certain agaropectin polymers (Kloareg and Quatrano 1988).

The joint presence of agar and other sulfated polysaccharides, such as carrageenans, although rare, is possible in certain species of red algae. This is particularly the case for the species *Grateloupia turuturu*, an alga consumed in Japan as a sea vegetable (Denis et al. 2010).

4.2.2.2. *Carrageenans*

Carrageenans are sulfated polysaccharides found in many species of red algae belonging mainly to the genera *Chondrus*, *Mastocarpus*, *Kappaphycus* and *Eucheuma* (Perez 1997). Carrageenans are a heterogeneous family of polysaccharides classified into three main subfamilies: kappa-carrageenans (κ), iota-carrageenans (ι) and lambda-carrageenans (λ) (see Figure 4.11).

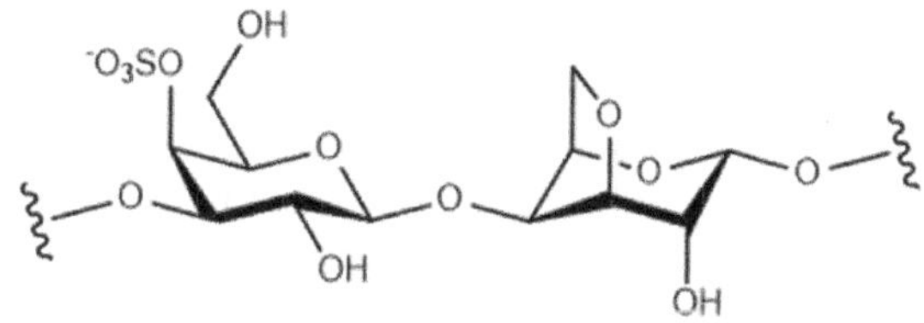

κ-carrageenans

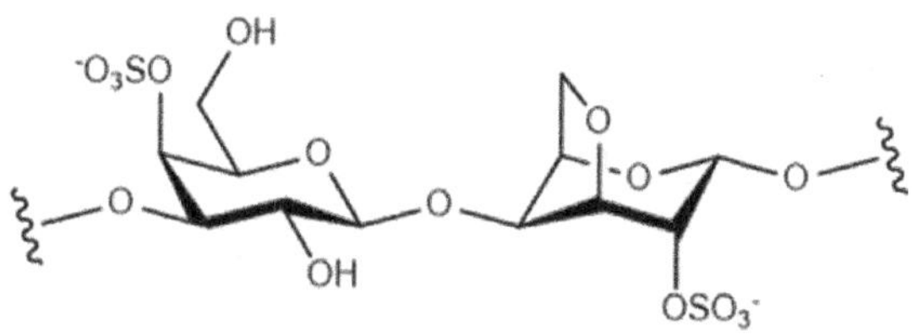

ι-carrageenans

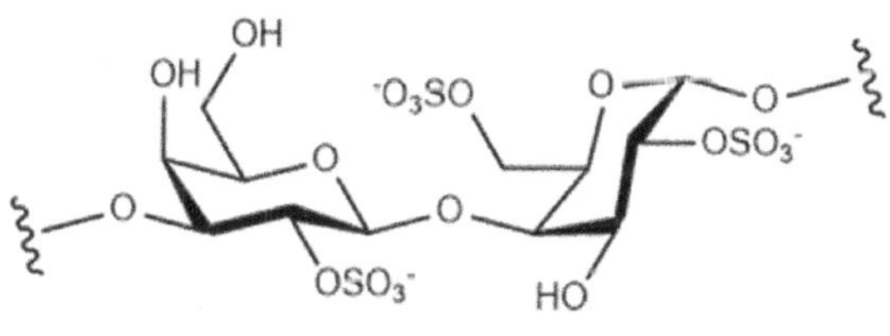

λ-carrageenans

Figure 4.11. *Structure of the different subfamilies of carrageenans*

κ-carrageenans are linear polymers resulting from the repeated linking of a D-galactose 4-sulfate molecule and a 3,6-anhydro-D-galactose molecule. ι-carrageenans are also linear polymers resulting from the repetition of a disaccharide composed of a D-galactose-4-sulfate and a 3,6-anhydro-D-galactose-2-sulfate. λ-carrageenans have the same linear structure but differ from the previous ones in their basic chain, which results from the combination of a D-galactose-2-sulfate with a D-galactose-2,6-sulfate.

The molecular masses reported for carrageenans vary according to the three molecular subfamilies but remain high. Thus, in the species *Euchema cottonii,* the molecular mass of ι-carrageenans is estimated at 738,000 Da (Viekbe et al. 1995). The κ-carrageenans obtained from the species *Euchema spinosum* are estimated at nearly 440,000 Da (Viekbe et al. 1995)

4.2.2.3. *Porphyrans*

Porphyrans are sulfated polysaccharides found mainly in algae of the genus *Porphyra,* which are used in human food under the name Nori (Chapter 2, section 2.2). Porphyran is a polysaccharide with a structure similar to that of agarose (Morrice et al. 1983). It consists of a chain of L-galactose-6-sulfate and D-galactose-6-methyl units (see Figure 4.12). Porphyran is the main water-soluble polysaccharide found in algae of the genus *Porphyra.* It can account for up to 40% of the alga's dry mass (Wang et al. 2023).

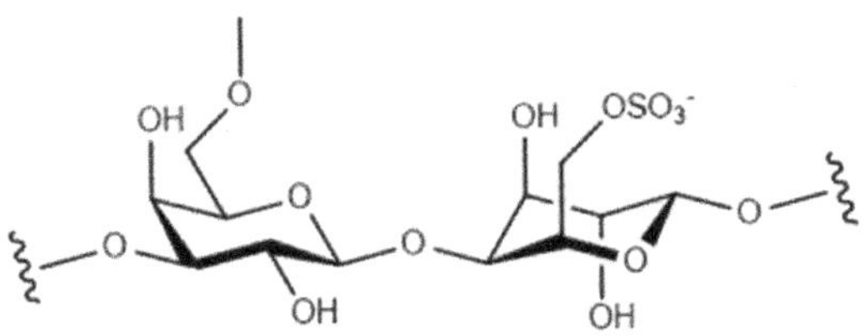

Figure 4.12. *Basic structure of porphyran*

4.2.3. *Carbohydrates in green algae (Chlorophyceae)*

Green algae of the genus *Ulva* have carbohydrate contents varying between 41 and 60% of dry matter (Darcy-Vrillon 1993). In Chlorophyceae, carbohydrates are also found mainly in the form of structural or storage polysaccharides.

Parietal fibrillar polysaccharides are mainly cellulose associated with a wide variety of hemicellulose (mannans, glucuronans, 1,3-β-glucans) (Stiger-Pouvreau et al. 2016). Matrix polysaccharides are branched polysaccharides with numerous acid and sulfate groups, better known as ulvans. Chlorophyceae also have reserve polysaccharides, such as fructin or a variety of starch (Stiger-Pouvreau et al. 2016).

4.2.3.1. *Ulvans*

Ulvans are considered to be the major constituents of green algae cell walls. They represent 8–29% of the alga's dried matter (Vera et al. 2011). They are complex polymers whose basic structure is a disaccharide unit repeated n times. This basic structural unit may consist of the combination of L-rhamnose-3-sulfate with D-guluronic acid (see Figure 4.13). The structure thus formed, or ulvabiouronic acid A, can combine with other ulvabiouronic acids to form highly heterogeneous ulvans depending on the nature of the substituents present on the other ulvabiouronic acids.

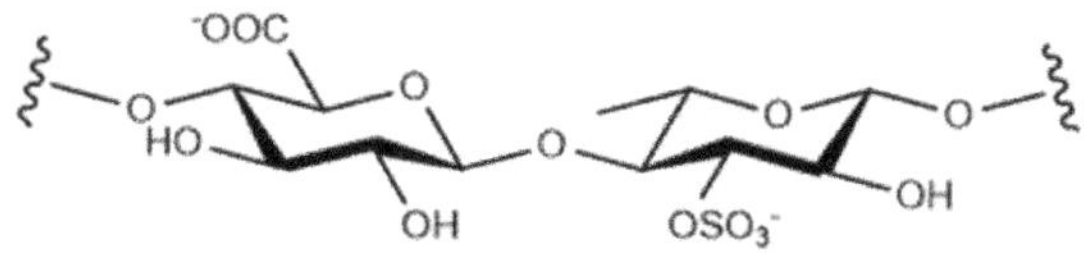

Figure 4.13. *Disaccharide structure of ulvabiouronic acid A*

4.3. Proteins

The protein content of marine algae varies greatly depending on the botanical family. In general, brown algae have relatively low protein content (3–15% of dry matter) (Fleurence 1999a).

Green algae, particularly those of the genus *Ulva*, often used as sea vegetables (Chapter 6, section 6.2), have protein contents ranging from 9 to 32% of the dry algal weight (Fleurence 2004; Fleurence et al. 2018). Red algae, on the other hand, are characterized by particularly high protein levels. These can reach up to 35% of dry matter in the species *Palmaria palmata* or Dulse and up to 47% in *Porphyra tenera* or Nori (Fujiwara-Arasaki et al. 1984; Fleurence 2004; Fleurence et al. 2012).

4.3.1. *Brown algae proteins (Phaeophyceae)*

The biochemical nature of brown algae proteins remains relatively unknown. Phaeophyceae have been studied mainly as an important source of unique polysaccharides (alginates, fucans) with numerous applications (Chapter 6, section 6.7). Proteins are present in low concentrations in brown algae compared to other algae species, and little research has been conducted on their biochemical characteristics. Another factor limiting the study of brown algae protein characteristics is the relative difficulty of extracting these compounds due to the high presence of anionic polysaccharides in the algal cell wall. In fact, parietal polysaccharides and their degree of cross-linking make the parietal structure highly resistant to conventional grinding techniques, thus limiting access to the protein fraction (Fleurence 1999b).

Brown algae proteins have been characterized mainly in terms of quantity, as well as in terms of quality by evaluating their amino acid composition. The protein content of brown algae varies depending on the species and season (Table 4.5) (Fleurence 2004; Fleurence et al. 2018).

Species	Protein content (% of DM)
Laminaria digitata	8–15
Ascophyllum nodosum	3–15
Undaria pinnatifida	11–24

Table 4.5. *Intra- and interspecific variations in the protein content in a few species of brown algae used in human food (Fleurence 2004)*

Among brown algae, the species *Undaria pinnatifida*, valued as a sea vegetable under the name Wakame, is particularly rich in protein, with a maximum protein value reported for this species of approximately 24% of dry matter.

The protein fraction of *U. pinnatifida* is characterized by the predominant presence of arginine (7.5% of the total protein amino acid fraction), aspartic acid (5.6%) and glutamic acid (5.1%) (Fleurence 2004; Fleurence et al. 2018). Leucine, lysine and phenylalanine, which are essential amino acids, represent 5.1%, 4.3% and 3.7% of the total protein amino acid fraction, respectively. These levels are lower than those reported for the same amino acids in the protein fraction of legumes such as soybeans (see Figure 4.14). However, the protein fraction of *Undaria pinnatifida* is a better source of phenylalanine and methionine than legume proteins such as soybeans (see Figure 4.14).

Undaria pinnatifida is characterized by a particularly high taurine content. This rare amino acid, available in free form, can be present at a concentration of 118 $mg.kg^{-1}$ of dry algal matter (Fleurence 2004).

4.3.2. *Red algae proteins (Rhodophyceae)*

Red algae are known to have the highest protein content. This is mainly due to the presence of a pigment protein called R- or B-phycoerythrin (Chapter 1, section 1.1.3). This purple-pink pigment can account for up to 12.2% of the protein fraction of the red alga *Palmaria palmata* (see Figure 4.15) (Galland-Irmouli et al. 2000). However, the R-phycoerythrin content is subject to significant seasonal variations, as is the case for all algal protein contents (see Figure 4.16).

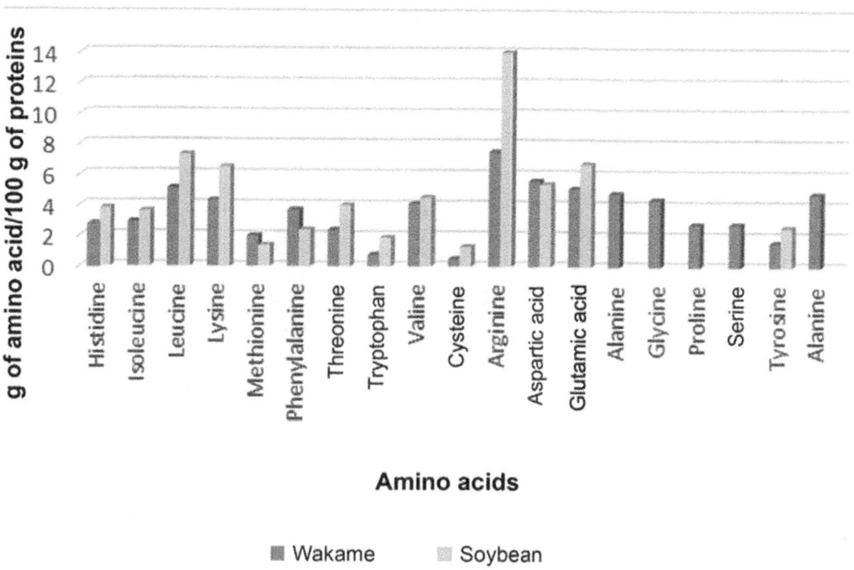

Figure 4.14. *Comparison of the amino acid composition of the protein fraction of the alga* Undaria pinnatifida *and a legume such as soybean (Fleurence et al. 2018)*

Figure 4.15. *Crude extract of R-phycoerythrin (RPE) obtained from the red alga* Palmaria palmata *(photo credit © Fleurence J. 2022)*

The protein content of certain red algae such as *Gracilaria gracilis* (Ogonori) or *Porphyra tenera* (Nori) can reach 45–47% of dry algal matter (Fujiwara-Arasaki et al. 1984; Dumay and Morançais 2016). These levels are comparable to those reported for soybean meal (44–48%), which is used in animal feed for its high protein content.

As with brown algae, Rhodophyceae have protein contents that vary depending on the species and season (Table 4.6) (Fleurence 2004; Denis et al. 2010; Francavilla et al. 2015; Rodrigues et al. 2015).

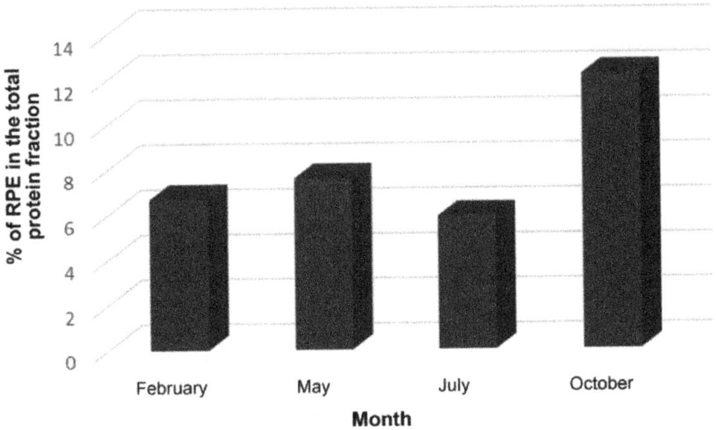

Figure 4.16. *Seasonal variations in R-phycoerythrin (RPE) content in the red alga* Palmaria palmata *(Galland-Irmouli et al. 2000)*

Species	Protein content (% of DM)
Gracilaria gracilis	31.00–45.00
Grateloupia turuturu	14.06–27.50
Porphyra tenera	33.00–47.00
Palmaria palmata	7.00–35.00

Table 4.6. *Intra- and interspecific variations in some red algae species used in human food (Morgan et al. 1980; Fleurence 2018)*

For the species *Palmaria palmata,* a red alga frequently used in Europe as a sea vegetable or ingredient, the highest protein contents are found in winter and early spring (see Figure 4.17).

Outside its natural environment, *Palmaria palmata* can also be cultivated in land basins. Under these conditions, it is possible to influence the protein content depending on the concentration of nitrogen added to the medium.

When the alga is cultivated in nitrogen-enriched water, the protein content observed is approximately 23.4% (see Figure 4.18). This content is close to the maximum protein content observed for this species at the end of winter and in early spring. In contrast, the same species produced in basins with nitrogen-depleted water shows a relatively low protein content of 8.1% of dry algal matter (Morgan et al. 1980).

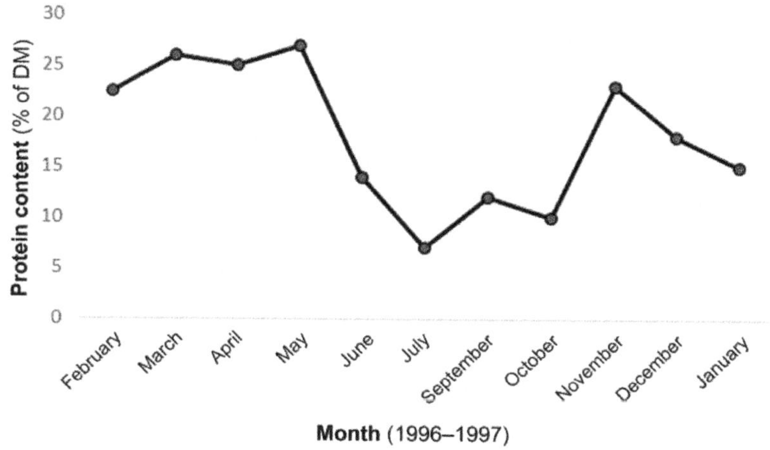

Figure 4.17. *Seasonal variation over a one-year period in the protein content of the red alga* Palmaria palmata *(Galland-Irmouli et al. 1999)*

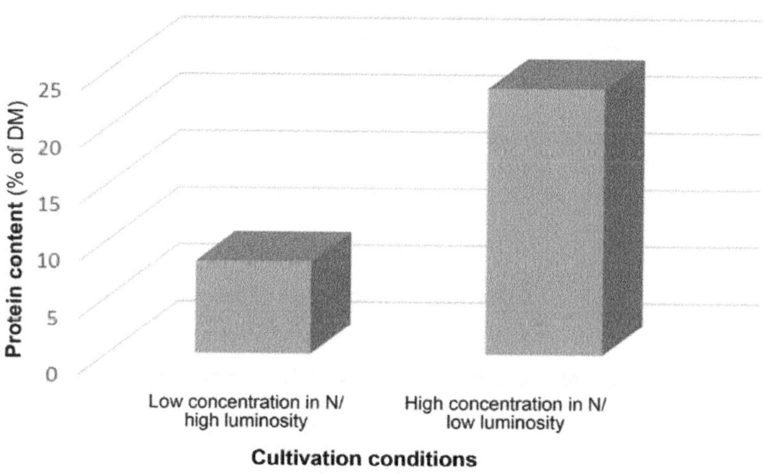

Figure 4.18. *Influence of cultivation conditions on the protein content in the alga* Palmaria palmata *(Morgan et al. 1980)*

The cultivation of the alga *Palmaria palmata* therefore appears to be an alternative to avoid the excessive seasonal variations in protein found in resources harvested from the natural environment.

Palmaria palmata, the main red alga consumed in Europe as a sea vegetable, has some unique characteristics in terms of the amino acid composition of its protein fraction.

Essential amino acids make up 36% of the total amino acid fraction. This includes free amino acids as well as those bound to proteins. Non-essential amino acids make up 64% of the total amino acid fraction of *Palmaria palmata* (see Figure 4.20).

Essential amino acids play a vital nutritional role as they are not synthesized by the body and must therefore be obtained from food. In humans, nine amino acids are considered essential. Cysteine and tyrosine are not strictly essential amino acids, as they can be synthesized from methionine and phenylalanine, respectively (see Figure 4.19).

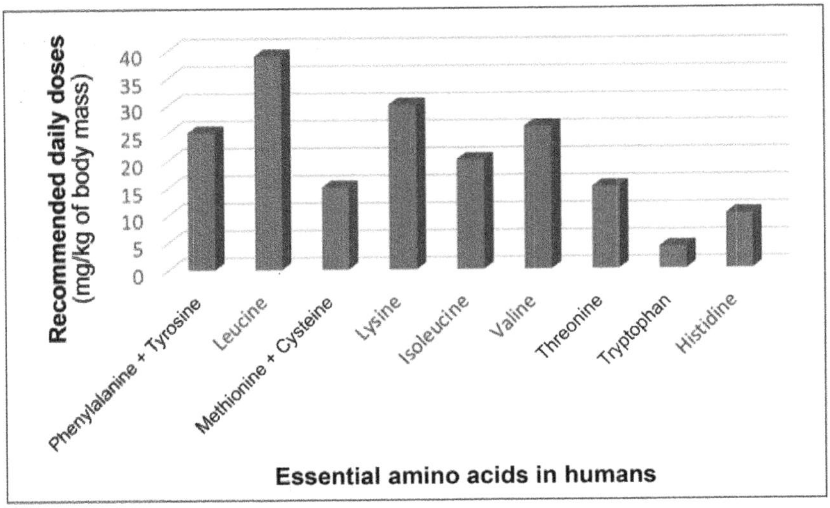

Figure 4.19. *Essential amino acids in humans and recommended daily allowances by the WHO (see: https://en.wikipedia.org/wiki/Essential_amino_acid)*

Box 4.2. *Essential amino acids in humans*

This ratio of essential to non-essential amino acids (36/64) can be compared to that observed for soybeans, which is approximately 35/65%.

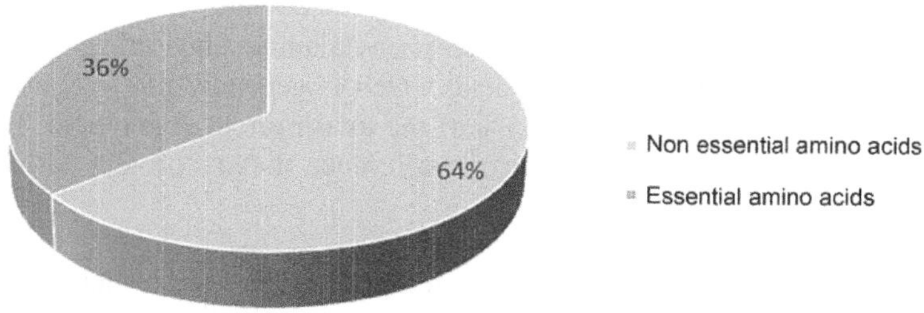

Figure 4.20. *Distribution between essential and non-essential amino acids in the protein fraction of* Palmaria palmata *(Fleurence 2018)*

Palmaria palmata therefore appears to be a source of essential amino acids (Box 4.2) comparable to soybeans, which are an important source of plant-based protein and amino acids for both animal and human nutrition.

	Palmaria palmata	*Porphyra tenera*	*Chondrus crispus*	*Grateloupia turuturu*	Soy	Ovalbumin
Histidine	0.5–1.2	1.4	0.9	1.8	3.8–4.0	4.1
Isoleucine	3.5–3.7	4.0	1.8	4.4	3.6	4.8
Leucine	5.9–7.1	8.7	2.9	6.3	7.3	6.2
Lysine	2.7–5.0	4.5	4.9	4.3	6.4–6.5	7.7
Methionine	2.7–4.5	1.1	0.5	2.0	1.2–1.4	3.1
Phenylalanine	4.4–5.3	3.9	1.5	3.7	2.4	4.1
Threonine	3.6–4.1	4.0	2.2	3.0	4.0	3.0
Tryptophan	3.0	1.3	–	0.7	1.6–1.9	1.0
Valine	5.9–6.1	6.4	–	4.9	4.5	5.4

Table 4.7. *Essential amino acid composition of some red algae used in human nutrition (in g of amino acids/100 g of protein) (Fleurence 2004; Dumay and Morançais 2016)*

The essential amino acids in *Palmaria palmata* are characterized by the predominant presence of leucine, lysine, phenylalanine and valine (Table 4.7). Methionine, a sulfur-containing amino acid, which is generally not very abundant in plant proteins, can constitute up to 4.5% of the total protein amino acid fraction of the algae. By comparison, ovalbumin, the reference dietary protein for the FAO, contains 3.1% of methionine.

Palmaria palmata is also characterized by a tryptophan content three times higher than that reported for ovalbumin and significantly higher than that recorded for soybeans (Table 4.7). With the exception of histidine and lysine, the essential amino acid composition of *Palmaria palmata* is equivalent to or higher than that of a nutritionally important legume such as soybeans (Table 4.7).

The alga *Grateloupia turuturu*, mainly consumed in Japan as a sea vegetable, has leucine, isoleucine and threonine contents similar to those of ovalbumin (Table 4.7) (Fujiwara-Arasaki et al. 1984).

Grateloupia turuturu is a red algae native to Japan. It was first described in 1941 by Yamada, hence its botanical name *Grateloupia turuturu* Yamada 1941. It was accidentally introduced into the Thau lagoon in the early 1980s with the importation of Japanese oyster spat (*Crassostrea gigas*). In the 1990s, it was found in Brittany. Currently, this species is also present on the European and American Atlantic coasts. *G. turuturu* is characterized by a high degree of polymorphism, which can cause it to be confused in certain circumstances with other red algae such as *Palmaria palmata.* Unlike the latter, *Grateloupia turuturu* is not yet authorized for food use in France (sea vegetable, ingredient). This species, also known as "Grateloupe", has many nutritional benefits for both humans and animals.

Box 4.3. Grateloupia turuturu, *a non-native red alga used as a sea vegetable in Japan*

4.3.3. *Green algae proteins (Chlorophyceae)*

Green algae also have protein contents that vary specifically and seasonally (Table 4.8).

Algae of the genus *Ulva*, used in human food, generally have a protein content of more than 20% of the dry matter of the algae (Fleurence 1999a). This can reach up to 32% of the dried algae, as is the case for *U. lactuca* (Fleurence 2004; Dumay and Morançais 2016). Other species such as *U. compressa* and *U. pertusa* have protein values of 26% of the dry weight of the algae (Patarra et al. 2011; Fleurence et al. 2018).

Species	Protein content (% DM)
Ulva lactuca	8.7–32.7
Ulva rigida	11.5–29.5
Ulva armoricana	18.0–24.0
Ulva pertusa	17.5–26.0
Ulva compressa	26.6
Enteromorpha prolifera	14.3
Enteromorpha linza	12.3
Codium fragile	15.6
Caulerpa racemosa	17.3
Caulerpa lentillifera	10.4
Caulerpa sertularioides	17.9

Table 4.8. *Interspecific and seasonal variations in the protein content of some green algae (Fleurence et al. 1999; Fleurence 2018)*

The species *U. armoricana*, an alga involved in the green tide phenomenon in Brittany, has a protein content that can represent up to 24% of the dry algal weight (Fleurence et al. 1999). Certain species of the genus *Monostroma* sp., consumed in Japan under the name "Hitoegusa", have an average protein content of 20% of dry algal matter (Nisizawa et al. 1987).

Some species of the genus *Enteromorpha* have been studied very little, and there is little or no information on seasonal variations in the protein content of these algae (Table 4.8). The figures available in the scientific literature show that the most commonly reported protein content ranges from 12 to 15% of the dry weight of the algae (Bruni and Stancher 1973; Dumay and Morançais 2016). Algae of the genus *Caulerpa*, mainly consumed in Asia and Oceania, also have protein contents ranging from 10.5 to 17.9% of dry matter (Matanjun et al. 2009; Komalavalli and Latitha 2015) (Table 4.8).

Most green algae, and more particularly algae of the genus *Ulva*, have an essential amino acid/non-essential amino acid ratio of approximately 35/65%. This ratio is comparable to that reported for the protein fraction of soybeans.

A species such as *U. armoricana* is characterized by a phenylalanine content of up to 7.1% of the total amino acid fraction contained in proteins (Fleurence et al.

1999). Its maximum leucine content (6.7%) is comparable to that reported for ovalbumin (6.2%) (Table 4.9).

The alga *U. pertusa* is characterized by a histidine content (4%) significantly higher than that recorded for other algae and is therefore an interesting source of this essential amino acid (Table 4.9).

U. rigida and *U. rotundata*, green algae found on French coasts and used in the product "sea lettuce" or "fisherman's salad", are distinguished by their threonine and valine contents, which is particularly interesting compared to that reported for soy or ovalbumin (Table 4.9).

Compared to other Ulva species, the algae *U. reticula* appears to be a potential source of methionine, with a content of 3% of the total protein amino acid fraction (Komalavalli and Latitha 2015). It is close to the content recorded for *P. palmata* (Table 4.7) and appears similar to that reported for ovalbumin.

Taurine, a rare free amino acid, already detected in brown algae of the genus *Undaria* or *Sargassum*, is also present in green algae. Its content is 18 mg.kg^{-1} of dry matter in the species *U. pertusa* and 20 mg.kg^{-1} of dry matter in the species *Enteromorpha linza* (Fleurence 2004).

	Ulva pertusa	*Ulva armoricana*	*Ulva rigida*	*Ulva rotundata*	Soy	Ovalbumin
Histidine	4.0	1.2–2.1	2.1–2.9	1.5–2.8	3.8–4.0	4.1
Isoleucine	3.5	2.3–3.6	4.5–6.3	4.1–7.6	3.6	4.8
Leucine	6.9	4.6–6.7	6.6–9.4	6.2–11.5	7.3	6.2
Lysine	4.5	3.5–4.4	4.0–5.6	4.9–9.0	6.4–6.5	7.7
Methionine	1.6	1.4–2.6	2.1–2.97	2.4–4.5	1.2–1.4	3.1
Phenylalanine	3.9	5.0–7.1	3.5–4.9	3.6–6.4	2.4	4.1
Threonine	3.1	4.5–6.8	6.8–9.6	6.0–9.6	4.0	3.0
Tryptophan	0.3	–	–	–	1.6–1.9	1.0
Valine	4.9	4.0–5.2	7.0–9.80	5.8–11.2	4.5	5.4

Table 4.9. *Essential amino acid composition of some* Ulva *species used or usable in human food (in g of amino acids/100 g of protein) (Fleurence et al. 1995, 1999; Fleurence 2018)*

Certain Chlorophyceae, such as algae of the genus *Caulerpa,* are commonly consumed in Southeast Asia and Oceania (Chapter 2, section 2.3). The species *Caulerpa racemosa,* well known in Fiji and Samoa under the name "green caviar", is characterized by particularly high lysine and leucine contents. These represent 7.0 and 7.7% of the total amino acid fraction, respectively (Komalavalli and Latitha 2015). This alga therefore appears to be as interesting a source of these two essential amino acids as soy or ovalbumin.

The presence of such high levels of leucine and lysine has also been confirmed in the species *Caulerpa sertularioides*. In the latter, the lysine concentration reaches 8.5% of the total amino acid fraction (Komalavalli and Latitha 2015), which is significantly higher than the lysine concentrations usually reported for most edible algae.

Algae of the genus *Caulerpa* are frequently consumed in the Philippines, Guinea and Japan as sea vegetables. The species used for this purpose are mainly *C. racemosa, C. lentillifera* and *C. sertularioides*.

The alga *C. lentillifera* is known as "sea grape" or umibudo in Japanese. It is traditionally used in Japanese cuisine and is part of the famous "Okinawa diet".

Figure 4.21. Caulerpa lentillifera *(Umibudo) (photo credit © https://commons.wikimedia.org/wiki/File:Umibudou_Detail.JPG)*

Box 4.4. Caulerpa *species frequently used for food*

4.4. Lipids and fatty acids

Marine algae generally have a low lipid content (0.5–7% of dry matter) (Nisizawa et al. 1987; Wielgosz-Collin et al. 2016). Some species, particularly red algae, may have extremely low levels depending on the season (0.1% of dry matter) (section 4.4.3).

Algae are therefore low-calorie foods due to their low lipid content. Relative to fresh weight, this is similar to that of certain land vegetables (see Figure 4.22).

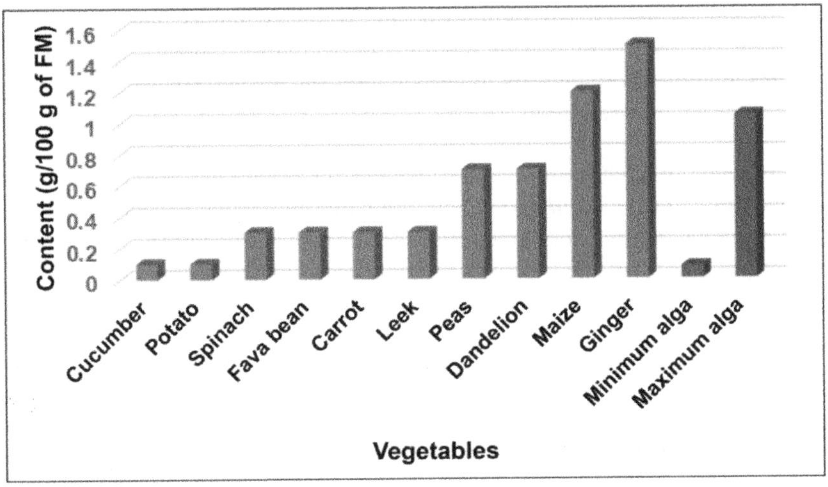

Figure 4.22. *Lipid content of some vegetables and comparison with the minimum and maximum values reported for algae*

Box 4.5. *Examples of lipid content in land vegetables and algae*

As with other compounds of nutritional interest such as proteins or carbohydrates, lipid content in algae is subject to interspecific and seasonal variations. However, seasonal variations in the algal lipid fraction are far from having been studied in all species of food interest. Knowledge in this area is still patchy. This is due to the relatively recent interest in algal lipids and the relatively limited number of studies on the subject. Algae also contain fatty acids, including long-chain polyunsaturated fatty acids such as omega-6 and omega-3 (arachidonic acid, eicosapentaenoic acid). In addition, marine algae contain unique sterols.

4.4.1. *Lipids in brown algae (Phaeophyceae)*

The lipid content of brown algae is subject to significant interspecific variations (1.4–6.5% of dry matter) (Fleurence et al. 1994) (Table 4.10).

Species	Lipids (% of DM)	Neutral lipids (% TL)	Phospholipids (% TL)	Glycolipids (% TL)
Laminaria digitata	1.5	59.2	11.5	29.3
Laminaria saccharina	3.2	64.7	10.9	24.4
Laminaria japonica	1.3	–	–	–
Laminaria religiosa	0.5	–	–	–
Undaria pinnatifida	6.5	15.1	33.9	51.0
Fucus vesiculosus	2.4	67.8	6.0	26.2

TL: total lipids

Table 4.10. *Lipid content and distribution of different lipid classes in some brown algae used for food (Fleurence et al. 1994; Fleurence 2018)*

The algae *Undaria pinnatifida* (Wakame), with a lipid content of 6.5% of dry weight, appears to be the richest species compared to other brown algae. This species has the particularity of preferentially storing lipids in fertile bases or sporophylls (see Figures 4.23 and 4.24) (Boulom et al. 2014).

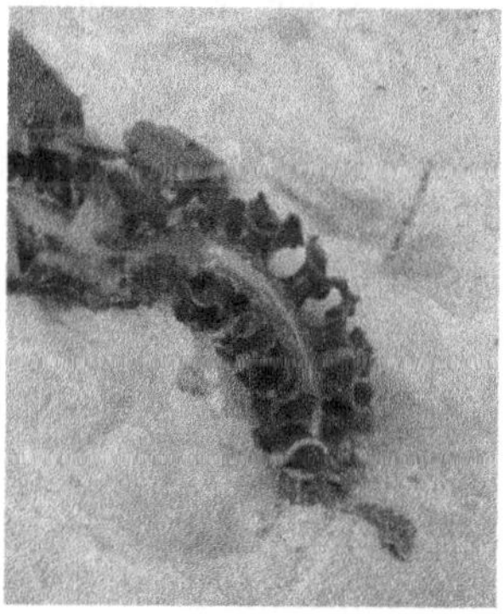

Figure 4.23. *Fertile bases (sporophylls) of the brown alga* Undaria pinnatifida *(photo credit © Fleurence J. 2023)*

However, the alga *Laminaria digitata* (Breton Kombu) with a lipid content of 1.5% of dry matter is classified as a low-lipid alga. The species *L. japonica*, consumed as Kombu, has a lipid fraction representing 1.3% of dry weight. *L. religiosa*, used in Japan to make "Hosome-Kombu", has an even lower lipid content, estimated at 0.5% of the alga's dried mass (Nisizawa et al. 1987).

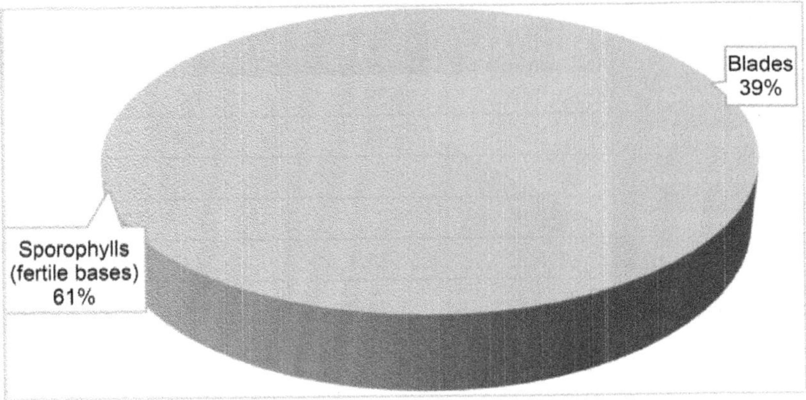

Figure 4.24. *Distribution of the lipid fraction in* Undaria pinnatifida *according to organ (blades, fertile bases) (Boulom et al. 2014)*

In addition to Laminariales of the genera *Undaria* and *Laminaria*, other brown algae used for food have also been studied for their lipid fractions. These include Fucales and, more specifically, the species *Fucus vesiculosus*. This alga, used in France under the name "*Goémon*" (Chapter 2, section 2.1), has a lipid content of 2.4% of dry algal matter (Table 4.10).

In brown algae, neutral lipids constitute the majority of the total lipid fraction (more than 60%) (Table 4.10). This does not apply to the species *U. pinnatifida*, where neutral lipids represent barely 15% of the total lipid fraction. Glycolipids and phospholipids are two to three times more abundant in this species than in most other brown algae used for food.

The lipid content of brown algae can also vary depending on environmental factors such as light intensity. This has been described in particular in the species *Costaria costata*, where the lipid content is 2.7% of dry matter when the algae are exposed to strong natural light and 6.7% when they are in a low-light environment (Hotimchenko 2002).

4.4.2. *Fatty acids and sterols in brown algae (Phaeophyceae)*

As in all algae, fatty acids are present in Phaeophyceae, either in free form or bound, in which case they are part of the composition of lipids.

Fatty acids, particularly polyunsaturated fatty acids with long carbon chains (C20), are of major nutritional interest for human health and nutrition (Newton 1996).

Free fatty acids and sterols are part of the "unsaponifiable" fraction, and the fatty acids contained in lipids are part of the "saponifiable" fraction. These three types of compounds are present in brown algae and in all other algal families.

In the algae *U. pinnatifida* (Wakame), the total fatty acid content (free fatty acids + bound fatty acids) is 16.6 $mg.g^{-1}$ of dry matter (Fleurence et al. 1994). This content is only 5.4 $mg.g^{-1}$ of dry matter in the species *L. digitata* (Breton Kombu) (Fleurence et al. 1994). These two brown algae, which belong to the order Laminariales, therefore have very different fatty acid contents and do not have the same nutritional value for the supply of fatty acids in human nutrition. Polyunsaturated fatty acids, such as arachidonic and eicosapentaenoic acids (C20:4n-6, C20:5n-3), known for their cardioprotective effects, are present in brown algae (Table 4.11).

Species	Arachidonic acid (C20:4n-6)	Eicosapentaenoic acid (C20:5n-3)
Undaria pinnatifida	17.5	8.2
Laminaria saccharina	13.7	6.2
Laminaria digitata	8.0	8.3
Fucus vesiculosus	8.7	4.7

Table 4.11. *Distribution between arachidonic acid and eicosapentaenoic acid in some brown algae consumed as sea vegetables (as a % of total fatty acids) (Fleurence et al. 1994; Fleurence 2018)*

With a content of arachidonic acid and eicosapentaenoic acid of 25.7% of the total fatty acid fraction, the algae *U. pinnatifida* appears to be a potential food source for these two polyunsaturated fatty acids. However, this finding must be qualified, as some species of brown algae have much higher contents. This is particularly true of the species *Cystoseira barbata*, whose cumulative content of these two compounds reaches 48.3% of the total fatty acid fraction (Dembitsky et al. 1990). However, this species is not among the algae commonly consumed.

Brown algae, such as *Undaria pinnatifida*, are potential sources of eicosapentaenoic acid, which are of interest for human nutrition. They can have levels similar to certain marine food sources such as fatty fish such as mackerel (*Scomber scombrus*).

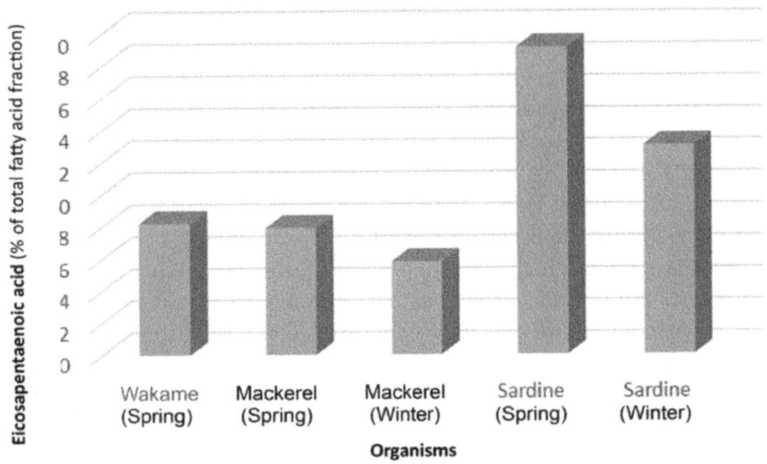

Figure 4.25. *Eicosapentaenoic acid content in different marine food sources*

Box 4.6. *Some examples of eicosapentaenoic acid content contained in the alga* Undaria pinnatifida *and various fatty fish (Fleurence et al. 1994; Bandarra et al. 1997; Guizani and Moujahed 2015)*

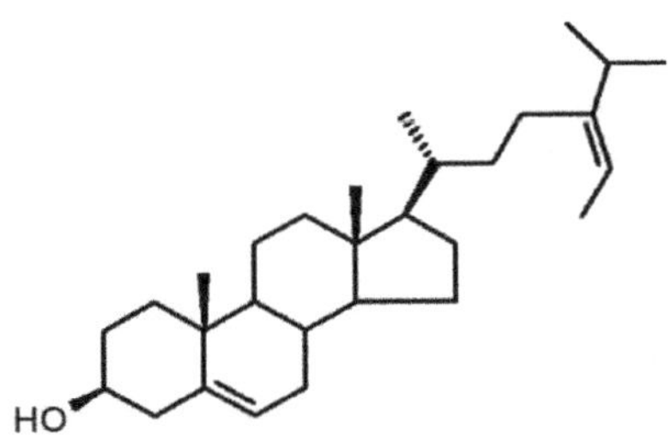

Figure 4.26. *Structure of a fucosterol molecule (Fleurence 2018)*

In addition to fatty acids, brown algae contain sterols, some of which have antioxidant and even antidiabetic properties. This is particularly true of fucosterol (see Figure 4.26), which is found in algae of the genus *Fucus*, particularly *F. vesiculosus* and *F. serratus*. This sterol is also present in *L. japonica* (Kombu), where it can account for 0.009% of the dry algal matter. Although very low, this content is much higher than those reported for red or green algae (Nisizawa et al. 1987). However, it is lower than that recorded for the brown alga *Hizikia fusiformis* (Hiziki) (0.05% of dry matter) (Nisizawa et al. 1987). In addition to fucosterol, the presence of 24-methylenecholesterol has also been reported in this alga. The latter sterol is present at a level of 0.007% of dry matter (see Figure 4.27).

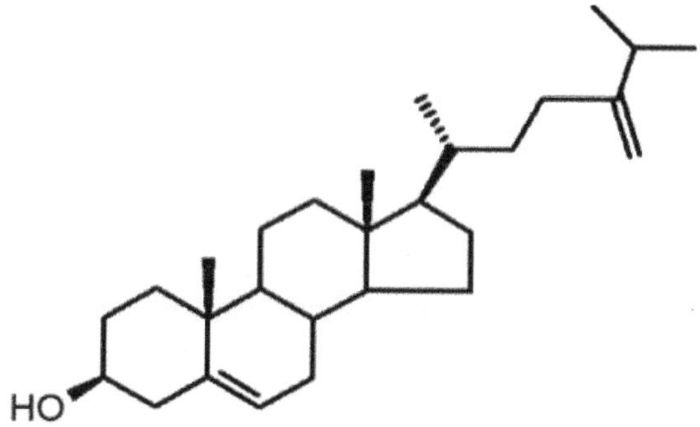

Figure 4.27. *Structure of a 24-methylenecolesterol molecule (Fleurence 2018)*

4.4.3. *Lipids in red algae (Rhodophyceae)*

The lipid content of red algae ranges from 0.1 to 6.2% of dry algal matter (Wielgosz-Collin et al. 2016). *Porphyra tenera*, an alga used in Japan to make "Hoshi-Nori", has a lipid content of 2.1% of the dried mass of the alga (Nisizawa et al. 1987). The species consumed in Europe have lipid contents ranging from 0.6 to 3.4% of the dry weight of the algae (Table 4.12) (Fleurence et al. 1994).

However, lipid content can vary depending on the season and certain environmental factors. Unfortunately, this variation has only been monitored in a limited number of species. In the species *Palmaria palmata*, variations in lipid content have been reported depending on the harvesting site and the sampling period (Morgan et al. 1980). Thus, depending on the season, the lipid content for this species ranges from 0.4 to 1.6% of dry matter. These values were obtained from algae collected on the French coast (Morgan et al. 1980). The same species collected in May on the coast of Nova Scotia (Halifax) shows a particularly high lipid content, estimated at 3.8% of dry matter (MacPherson and Young 1949).

Species	Total lipid (% of DM)	Neutral lipids (% of TL)	Phospholipids (% of TL)	Glycolipids (% of TL)
Porphyra umbilicalis	3.4	18.9	25.2	55.9
Chondrus crispus	0.6	19.8	23.7	56.5
Palmaria palmata	1.6	23.0	15.4	61.6
Gracilaria verrucosa	1.4	7.3	19.6	73.1
Grateloupia turuturu	3.6	32.9	21.9	45.1

TL: total lipids

Table 4.12. *Lipid content and distribution of different lipid classes in some red algae used for food in Europe or Japan (Fleurence et al. 1994; Fleurence 2018)*

Other harvests carried out on the Icelandic coast (Munda 1972) and Grand Manan Island (New Brunswick) reported lipid contents varying between 0.6 and 1.7% of the dry weight of the algae (Idler and Wiseman 1970).

Seasonal variation in lipid content has been studied in particular in the species *Grateloupia turuturu*. This red alga, introduced to French coasts, is consumed in Japan as a sea vegetable. The lipid content of this species varies between 3.3 and 4.1% of dry matter throughout the year (Kendel et al. 2013). The percentage of glycolipids varies little depending on the period (42.3–46.8% of the total lipid fraction). However, strong seasonal variations are associated with the other two classes of lipids (neutral lipids, phospholipids). The neutral lipid content increases from 20.1% in summer to 41.8% in winter. A similar trend is observed for phospholipids, whose content is 11.8% in winter compared to 33.4% in summer (Kendel et al. 2013).

The lipid content of *G. turuturu* also appears to be sensitive to environmental factors such as light intensity. This phenomenon, which has already been described in certain brown algae, therefore also applies to certain red algae. Thus, the lipid content of algae in a highly lit environment can fall to 0.7% of the dried mass (Hotimchenko 2002).

4.4.4. *Fatty acids and sterols in red algae (Rhodophyceae)*

Red algae are characterized by a high proportion of long-chain polyunsaturated fatty acids (C20).

Palmaria palmata and *Porphyra umbilicalis* are interesting sources of eicosapentaenoic acid (Fleurence et al. 1994). This polyunsaturated fatty acid accounts for 47–48% of the total fatty acid fraction in the two species mentioned above (Table 4.13). *Chondrus crispus,* another edible alga, is distinguished by a high proportion of arachidonic acid (22.5% of the total fatty acid fraction).

Species	Arachidonic acid (C20:4*n*-6)	Eicosapentaenoic acid (C20:5*n*-3)
Palmaria palmata	1.4	46.6
Porphyra umbilicalis	10.9	48.0
Gracilaria verrucosa	5.3	25.8
Chondrus crispus	22.5	18.7
Grateloupia turuturu	8.0	13.5
Grateloupia divaricata	28.3	14.8

Table 4.13. *Distribution of arachidonic acid and eicosapentaenoic acid in some red algae consumed as sea vegetables (as a percentage of total fatty acids) (Khotimchenko and Vaskovsky 1990; Fleurence et al. 1994; Schmid et al. 2014)*

Among edible red algae, *Palmaria palmata* appears to be an interesting source of eicosapentaenoic acid. Based on dry matter, this alga can contain 0.4–0.6% of this polyunsaturated fatty acid (Schmid et al. 2014). The eicosapentaenoic acid content in *P. palmata* also appears to be subject to seasonal fluctuations, but these are small (+/–5%) when expressed as a percentage of the total fatty acid fraction between winter and summer (Schmid et al. 2014). In contrast, a species such as *G. divaricata* would seem to be more suitable as a food source of arachidonic acid (Table 4.13) (Khotimchenko and Vaskovsky 1990).

There is little information available on the nature of sterols and phytosterols present in Rhodophyceae. However, there are fairly comprehensive studies on the chemical nature of sterols and seasonal variation in their content in the algae *P. palmata* and *G. turuturu*.

In *P. palmata*, demosterol is the main sterol, accounting for 30–97% of total sterols contained in the alga (see Figure 4.28). Cholesterol is less abundant, accounting for 2–49% of the sterols present, depending on the season (Idler and Wiseman 1970).

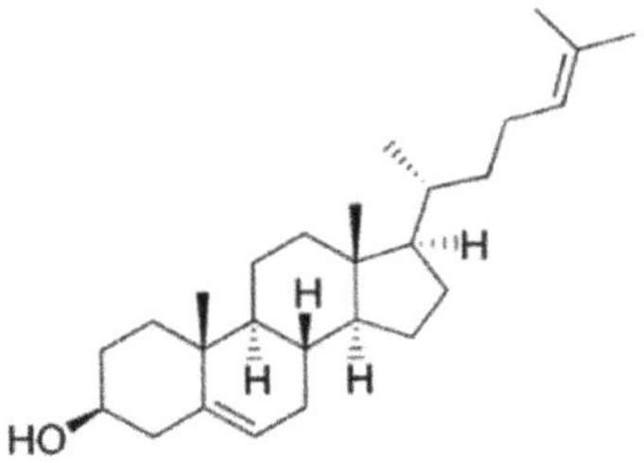

Figure 4.28. *Demosterol molecule (Fleurence 2018)*

In *G. turuturu*, sterols account for 47% of the unsaponifiable fraction, with the remainder consisting of free fatty acids (Kendel et al. 2013). Cholesterol alone accounts for nearly 38% of this fraction, followed by lathosterol (1.9%), brassicasterol (1.1%) and finally fucosterol (1%).

4.4.5. *Lipids in green algae (Chlorophyceae)*

As with other algae, the lipid content of green algae is subject to interspecific and seasonal fluctuations (Table 4.14).

Species	Lipids (% of DM)	Neutral lipids (% TL)	Phospholipids (% TL)	Glycolipids (% TL)
Ulva rotundata (1)	2.0	46.0	16.1	37.9
Ulva lactuca (2)	4.1	–	–	–
Ulva reticula (3)	1.9	–	–	–
Enteromorpha intestinalis (1)	1.9	30.3	14.7	55.0
Caulerpa racemosa (3)	2.2	–	–	–
Caulerpa lentillifera (4)	1.1	–	–	–

TL: total lipids

(1) (Fleurence et al. 1994). (2) (Khairy and El-Shafay 2013). (3) (Hong et al. 2007). (4) (Matanju et al. 2009).

Table 4.14. *Lipid content of some green algae used in human food*

In the family Ulvaceae (Ulves, Enteromorphs), green algae commonly consumed in human food, the lipid content represents 0.7–5.1% of dry matter (Wielgosz-Collin et al. 2016). Algae of the genus *Monostroma*, used in Japan in the manufacture of "Hitoegusa", have an average lipid content of 1.2% of dry matter (Nisizawa et al. 1987).

Most species are also subject to seasonal variations in lipid content. For example, the lipid content of the alga *Ulva lactuca* increases from 3.1% of dry matter in the fall to 4.1% in spring (+25%) (Khairy and El-Shafay 2013).

However, there is little information on the nature of these lipids in green algae and their distribution by chemical family. Existing studies show that of the 24 species of green algae studied, the class of neutral lipids appears to be predominant (45.7%), followed by glycolipids (39.5%) and phospholipids (14.8%) (Wielgosz-Collin et al. 2016).

4.4.6. *Fatty acids and sterols in green algae (Chlorophyceae)*

Total fatty acids represent 0.5–0.7% of the dry matter of algae such as *Ulva rotundata* and *Enteromorpha intestinalis* (Fleurence et al. 1994).

Species	Arachidonic acid (C20:4*n*-6)	Eicosapentaenoic acid (C20:5*n*-3)
Ulva lactuca (1)	1.1	3.9
Ulva rotundata (2)	0.3	1.0
Enteromorpha intestinalis (2)	0.7	1.0
Caulerpa lentillifera (3)	–	0.9

(1) (Schmid et al. 2014). (2) (Fleurence et al. 1994). (3) (Matanjun et al. 2009).

Table 4.15. *Distribution between arachidonic acid and eicosapentaenoic acid in some green algae consumed as sea vegetables (as a percentage of total fatty acids)*

By way of comparison, the brown alga *Undaria pinnatifida* and the red alga *Porphyra umbilicalis* have fatty acid levels of 1.7% and 1.3% of dry algae matter, respectively. The levels of arachidonic acid and eicosapentaenoic acid found in green algae (Table 4.15) are generally much lower than those reported for brown or red algae (Tables 4.11 and 4.13). These levels also fluctuate seasonally, sometimes significantly (–50%). This is particularly the case for the species *U. lactuca*, whose eicosapentaenoic acid content falls from 3.9% of total fatty acids in June to 1.6% in November (Schmid 2014). In the species *U. armoricana,* various sterols have been identified (Kendel et al. 2015). These are mainly cholesterol (35% of the unsaponifiable fraction) and isofucosterol (25%). The presence of brassicasterol (3%) and fucosterol (1%) has also been reported in this alga.

	Himanthalia elongata	*Laminaria digitata*	*Undaria pinnatifida*	*Chondrus crispus*	*Palmaria palmata*	*Ulva* sp.	Green or brown lentils	Brown rice
Na	600.6	624.6	448.6	1572.5	255.2	340.0	12.0	28.0
K	1351.4	2013.2	62.4	827.5	1169.6	245.0	940.0	1160
Ca	30.0	364.7	112.3	373.8	148.8	325.0	71.0	110.0
Mg	90.1	403.5	78.7	573.8	97.6	465.0	110.0	520.0
Fe	5.0	45.6	3.9	6.6	12.8	15.3	11.1	12.9
Zn	1.7	1.6	0.3	–	0.3	0.9	3.9	16.2
Cu	0.1	0.3	0.2	0.1	0.4	0.3	1.0	1.3
I	10.7	70.0	3.9	6.1	10.2	1.6	–	–

Table 4.16. *Examples of the mineral composition of some edible algae (mg/100 g wet weight) (MacArtain et al. 2007)*

4.5. Minerals

Algae are marine organisms with a high capacity to accumulate minerals dissolved in the aquatic environment. This characteristic is both an advantage and a disadvantage from a nutritional point of view. The advantage lies in the fact that algae concentrate certain elements, including trace elements such as calcium, magnesium and iron, at concentrations equivalent to or much higher than those found in terrestrial plant foods (Table 4.16). For example, algae of the genus *Ulva* have calcium levels that are three times higher than those of brown rice and almost five times higher than those of legumes such as green or brown lentils (Table 4.16) (MacArtain et al. 2007). The species *L. digitata, C. crispus* and *Ulva* spp. also stand out from other edible algae due to their magnesium content, which ranges from 403 to 575 mg/100 g of fresh product. These concentrations are equivalent to those recorded for brown rice (525 mg/100 g of fresh product). However, they remain significantly higher than those reported for lentils (110 mg/100 g of fresh product) and spinach (54 mg/100 g of fresh product) (McArtain et al. 2007). The algae *L. digitata*, *Palmaria palmata* and *Ulva* sp. appear to be sources of dietary iron equivalent or superior to certain cereals and legumes (Table 4.16). In terms of minerals of nutritional interest in human nutrition (Ca, Mg, Fe), algae of the genus *Laminaria* and *Ulva* seem to be the most interesting for dietary intake of these elements.

However, this observation on the nutritional benefits of algae consumption must be tempered by the high sodium content of these organisms. Most algae consumed have a sodium content ranging from 255 mg/100 g of fresh product to 1,572 mg/100 g of fresh product.

These levels are well above those found in vegetables or fruit (1–140 mg/100 g of fresh product).

The ability of algae to concentrate minerals dissolved in the environment also has a downside. They can accumulate heavy metals or toxic minerals present in the marine environment (Pb, Cd, Hg, As) in a non-specific manner.

> FRENCH REGULATIONS ON EDIBLE ALGAE (CHAPTER 7, SECTION 7.3.1).–
>
> This last aspect led to the introduction in France of regulations applicable to edible algae, defining toxicological mineral criteria to be met.

Most species consumed in Europe have toxic mineral content levels that comply with established criteria. Brown algae, which are more sensitive than other algae to heavy metal accumulation, generally have very low levels of lead and cadmium. However, the species *Undaria pinnatifida* (Wakame) has cadmium levels ranging from 0.28 to 1.38 $mg.kg^{-1}$ of dry matter (Fleurence and Le Cœur 1994; Jurkovic et al. 1995). Similarly, its lead content can sometimes exceed the maximum permitted concentration of 5 $mg.kg^{-1}$ of dry matter (Chapter 7, section 7.3.1) (Fleurence and Le Cœur 1994). In Japan, this alga is described as having a lead content ranging from 0.51 to 1.51 $mg.kg^{-1}$ of dry matter (Tanaka et al. 1973), which is well below the regulatory criterion applied for this element in France.

These seemingly contradictory pieces of information simply highlight the fluctuating nature of the mineral composition of an algal species, depending on the environment and therefore the harvesting site. This influence of the harvesting site on the mineral composition of algae, particularly brown algae, has also been demonstrated for iodine. This element, which is beneficial at low concentrations, can be toxic at high concentrations and is also subject to a regulatory limit (Chapter 7, section 7.3.1).

Brown algae of the genus *Laminaria* have very high iodine levels compared to other algae species, all families combined. In fact, the iodine content in Laminaria is often higher than 4,000 mg.kg^{-1} of dry matter and can reach 10,000 mg.kg^{-1} of dry matter, as in the case of *L. digitata* (Mabeau and Fleurence 1993). This iodine content, which fluctuates depending on the season, is not the same as that found in processed products available for consumption. In these products, it is generally less than 5,000 mg.kg^{-1} of dry matter. It can also vary depending on the geographical location and traditional preparation methods (Table 4.17).

Sample	1	2	3	4	5	6	7	8	9	10
Origin	Korea	Japan	Japan	Japan	Japan	Taiwan	Japan	China	Taiwan	Taiwan
Iodine (mg.kg^{-1} of DM)	4921.3	303.2	3297.7	1764.4	2250.8	4384.8	241.0	2236.2	3541.6	2291.6

Table 4.17. *Variation in iodine concentration in Kombu (*Saccharina japonica/Laminaria japonica*) according to the product's geographical origin (Yeh et al. 2014)*

In general, the mineral composition of algae is subject to interspecific, seasonal and geographical variations. In this context, it is very difficult to establish an average composition. Despite this, different species can be distinguished by their relative richness in one or more particular elements.

Laminaria are characterized by their high iodine content and certain minerals of interest such as calcium, potassium and magnesium (Table 4.16). Ulva, like Enteromorpha, appear to be potential sources of dietary calcium and magnesium.

LIMITS TO THE USE OF ALGAE AS A DIETARY MINERAL SUPPLEMENT

The wide variability in the mineral composition of algae is a factor that must be taken into account, particularly in the context of agri-food use.

4.6. Vitamins

The vitamin composition of marine algae has been the subject of numerous publications. However, the information collected on this subject is far from easy to use, particularly when it comes to inter- or intra-species comparisons. This is mainly due to the fact that the analyses were sometimes carried out on unprocessed resources, as well as on processed products. In the case of processed products, the analyses carried out give vitamin contents that are probably lower than those of the raw material. This is because most of the processes used involve moderate or no heating, which in principle denatures many vitamins. The variability of the analytical methods used (colorimetric analyses, HPLC analyses) combined with the diversity in the expression of results (moist matter or dry matter) also often makes interspecific comparisons difficult.

4.6.1. *Vitamins in unprocessed algae*

The vitamins most sought after in algae are those with major nutritional benefits. These are mainly vitamins A, C, D and E, as well as B vitamins, particularly vitamin B_{12}. The latter is found in low quantities in plants, and vegetarian, and more recently vegan seaweed eaters, face a potential vitamin B_{12} deficiency due to their dietary choices.

B vitamins have therefore been sought in many species of edible algae (Hong et al. 2007). One of the species richest in these vitamins is the alga *Ulva reticula* (12.7 mg.kg^{-1} of wet weight) (Table 4.18). This species is also characterized by a particularly high vitamin C content (145.6 mg.kg^{-1} of wet weight). The species *U. lactuca,* commonly consumed under the name "sea lettuce" has a vitamin C content of 100 mg.kg^{-1} of wet weight. This confirms the value of Ulvales as a potential source of vitamin C. However, some red algae also have high levels of vitamin C. This is particularly true of *Gracilaria changii*, whose vitamin C concentration can reach 285 mg.kg^{-1} of wet algal matter (Norziah and Ching 2000).

These vitamin C levels reported for Ulvales or certain Gracilaria species are much higher than those commonly accepted for peas, carrots, lettuce and tomatoes (Norziah and Ching 2000). They are similar to those recorded for vegetables such as broccoli (85 mg.kg^{-1} of wet matter) and red chili peppers (175 mg.kg^{-1} of wet matter).

Ulvales also stand out for their relative richness in vitamin A; the content of this vitamin can reach 21.2 mg.kg^{-1} of wet matter in *U. reticula* (Table 4.18).

While vitamins B_1, B_2, B_3, B_6, B_8 and B_9 have been easily identified in algae, vitamin B_{12} has been more difficult to detect. It has been identified in the red alga *Porphyra yezoensis* (Watanabe et al. 2000). The edible parts of this alga (lamellae or "leaves") contain 0.51 $mg.kg^{-1}$ of vitamin B_{12} in wet weight. The presence of this vitamin has also been reported in the red alga *P. palmata* (Morgan et al. 1980). In this species, its content varies between 0.009 and 0.24 $mg.kg^{-1}$ of dry matter.

Food	Vitamin A	Vitamin C	Vitamin E	B vitamins	B_{12}
Ulva reticulata (1)	2.12	145.6	–	12.7	–
Caulerpa racemosa (1)	2.16	136.8	–	10.5	–
Caulerpa lentillifera (2)	–	10.0	22.0	–	–
Gracilaria tenuistipitata (1)	2.10	75.30	–	7.3	–
Gracilaria changii (3)	–	285	–	–	–
Porphyra yezoensis (4)	–	–	–	–	0.51
Palmaria palmata (5)	54.0	38.8	–	8.1	0.03
Peas (5)	8.30	26.0	–	25.6	–
Carrot (5)	144.0	43.0	–	10.2	–
Orange (5)	–	490	–	4.3	–

(1) (Hong et al. 2007). (2) (Ratana-Arpon and Chirapart 2006). (3) (Norziah and Ching 2000). (4) (Watanabe et al. 2000). (5) (Morgan et al. 1980).

Table 4.18. *Comparison of vitamins in some edible algae and some vegetables or fruits ($mg.kg^{-1}$ of wet matter)*

4.6.2. *Vitamins in processed algae*

Algae processing involves a wide variety of methods, including drying, based on thermal drying, salting, freezing, appertization and even lactofermentation (Chapter 5). The influence of these processes on the vitamin content of the raw material used has rarely been studied. Nevertheless, the vitamin composition of seaweed-based products has often been analyzed, particularly with a view to promoting the nutritional value of seaweed in human nutrition.

The vitamin composition of the product “Kombu” (dried *Laminaria japonica*) has been analyzed in particular and compared with that of other vegetables (Nisizawa et al. 1987) (Table 4.19).

Vegetable	Vitamin C	Vitamin B_1	Vitamin B_2	Vitamin B_{12}	Provitamin A
Kombu	280	5.3	4.1	0.003	11
Spinach	4500	7.0	13.0	–	360

Table 4.19. *Comparison of the content of certain vitamins in Kombu and spinach (mg.kg^{-1} of dry matter) (Nisizawa et al. 1987)*

For Wakame (*Undaria pinnatifida*), there is more information available on the vitamin content of the raw material and derived products (Nisizawa et al. 1987). This information allows us to assess the influence of processing methods on the preservation or loss of certain vitamins (Table 4.20).

Thus, seaweed dried in the open air and in full sunlight does not appear to suffer any loss of vitamins, particularly vitamin C. The same is true for a derivative product, Haiboshi Wakame, when it is produced using a drying process such as open-air drying. However, the same product produced using a process that includes a washing step has a very low vitamin content, with some vitamins, such as vitamin C, completely absent.

Similarly, blanching the seaweed appears to be a process that denatures vitamin C, which is absent from blanched and salted Wakame (Table 4.20). This absence is also found in the product “cut Wakame”. This can be explained by the intermediate steps used in the production of this product, which involve pre-boiling the seaweed in brine, followed by washing with fresh water (Nisizawa et al. 1987).

More recently, a study conducted on the brown alga *Himanthalia elongata* (sea spaghetti) highlighted the influence of cooking processes on the product’s vitamin content (Amorin-Carrilho et al. 2014). Steaming or boiling completely denatures the vitamin C that was initially present in the dried seaweed at a content of 20.7 mg.kg^{-1} of dry matter.

Product	Vitamin C	Vitamin B_1	Vitamin B_2	Vitamin B_3 (niacin)	Provitamin A
Raw Wakame	150	0.7	1.8	9	14
Air-dried Wakame	150	3	11.5	80	33
Haiboshi Wakame	270	0.7	0.2	9	14
Washed Haiboshi Wakame	0	0	0.3	0	0.5
Blanched and salted Wakame	0	0.3	0.7	2	8.4
Chopped Wakame	0	1.1	0.7	1.3	56.3

Table 4.20. *Vitamin composition of some processed products based on* Undaria pinnatifida *(Wakame) (in mg.kg^{-1} of dry matter) (Nisizawa et al. 1987)*

The vitamin composition of Nori (*Porphyra yezoensis*, *Porphyra tenera*) and more specifically one of its derivatives, Hoshi-Nori, has been the subject of several studies. Hoshi-Nori is distinguished by its particularly high vitamin C content (1,125 mg.kg^{-1} of dry matter or 171 mg.kg^{-1} of wet product) (Nisizawa et al. 1987) and a fairly high concentration of B vitamins (B_1, B_2, B_3, B_6, B_{12}) (171.8 mg.kg^{-1} of dry matter). In dried Nori flakes, the vitamin B_{12} content can reach a concentration of 0.25 mg.kg^{-1} of dry matter.

This essential vitamin for health (O'Leary and Samman 2010) is also present in processed green algae products. For example, in dried flakes of *Enteromorpha* sp. used in the composition of the product Aonori, the reported vitamin B_{12} concentration is 0.69 mg.kg^{-1} of dry matter (Watanabe et al. 1999). Like the raw material, products derived from green algae processing are characterized by high vitamin C content (Table 4.21). This is particularly true for Aonori, which is generally a mixture of three algae (*Ulva* spp., *Enteromorpha* spp. and *Monostroma* spp.).

Product	Vitamin C	Vitamin B_1	Vitamin B_2	Vitamin B_3
Sea lettuce (*Ulva* sp.)	120	0.8	5.7	118
"Green laver" (*Enteromorpha* sp.)	432	6	20.5	65
Hitoegusa (*Monostroma* spp.)	540	4.3	13.3	35

Table 4.21. *Vitamin content of products used in the composition of Aonori (in mg.kg^{-1} of dry matter) (Nisizawa et al. 1987)*

4.7. Nutritional quality

Determining the biochemical composition of algae is one way of assessing their nutritional value in terms of energy intake. However, the nutritional quality of any food depends on the bioavailability of the nutrients it contains.

In the case of algae, certain components are not digested by the body and others have reduced digestibility due to the presence of antinutritional factors. The main factors implicated are fibers, which are indigestible components of the plant fraction (Rupérez and Toledano 2003).

4.7.1. *Dietary fiber*

Many algae polysaccharides behave like dietary fiber. Two types of fiber are characterized: soluble fiber and insoluble fiber (Lahaye 1991). Fiber is present in all edible algae, with soluble fiber representing the majority (Table 4.22).

This is a characteristic of algae and, as such, they appear to be an interesting source of soluble fiber for human consumption. In fact, only beet pulp has a soluble fiber content (25%) close to that of certain algae (*U. lactuca, Himanthalia elongata*), with other terrestrial plant sources consisting mainly of insoluble fiber.

Product	Soluble fiber	Insoluble fiber	Total fiber
Undaria pinnatifida (brown alga)	30.0	5.3	35.3
Hizikia fusiformis (brown alga)	32.9	16.3	49.2

Himanthalia elongata (brown alga)	25.7	7.0	32.7
Eisenia bicyclis (brown alga)	59.7	14.9	74.6
Ulva lactuca (green alga)	21.3	16.8	38.1
Enteromorpha sp. (green alga)	17.2	16.2	33.4
Porphyra tenera (red alga)	17.9	16.8	34.7
Apple	5.9	8.3	14.2
Cabbage	16.8	17.5	34.3
Wheat bran	8.0	77.0	85.0
Beet pulp	25.0	50.0	75.0

Table 4.22. *Fiber composition of some edible algae (as a % of dry matter) and terrestrial plant foods (Lahaye 1991; Mabeau and Fleurence 1993)*

Algal fibers, whether soluble or insoluble, can be broken down to varying degrees by human intestinal flora, but are poorly metabolized (Mabeau and Fleurence 1993). However, they are of definite interest from a physiological point of view. Insoluble fibers are poorly fermented but have a significant laxative effect. They also help to delay gastric emptying. Through this mechanism, they induce satiety more quickly and act as a natural appetite suppressant. Soluble algal fibers are more easily broken down. They help to reduce blood cholesterol (–18%) and promote the development of intestinal microbiota (Peñalver et al. 2020).

The nutritional value of algal polysaccharides therefore lies in their behavior as dietary fibers that facilitate intestinal transit and contribute to the health of the intestinal microbiota.

ALGAL FIBERS, ANTINUTRITIONAL FACTORS THAT LIMIT PROTEIN DIGESTION.

Algal fibers also act as antinutritional factors. By binding to proteins or inhibiting the action of digestive proteases (trypsin, chymotrypsin, pepsin), they limit the digestibility of the algal protein fraction.

The presence of factors limiting the digestibility of algal proteins has been described in both brown and red algae. In Phaeophyceae such as *L. japonica* (*Saccharina japonica*) (Kombu), *Undaria pinnatifida* (Wakame) and *Hizikia fusiformis* (Hijiki), the main antinutritional factors implicated are soluble fibers (Horie et al. 1995). The in vivo digestibility of the protein diet (20% casein) given to young rats is significantly affected (–11%) when 5% soluble fiber from *U. pinnatifida* is added to the basic diet.

The mode of action of soluble brown algae fibers appears to be a mechanism that inhibits certain digestive proteases such as pepsin (Table 4.23). This inhibition differs depending on the biological origin of the fibers and appears to be dependent on the biochemical nature of the polysaccharides involved. *L. japonica* fibers are the most effective at inhibiting pepsin, since 0.75% (mass/volume) of soluble fibers inhibit almost all pepsin activity (99.1%). In contrast, soluble fibers from *U. pinnatifida* have a lower inhibitory effect than those from *L. japonica* or *H. fusiformis* (Table 4.23).

Product	Percentage of soluble fiber added	Percentage inhibition of pepsin
Laminaria japonica (Kombu)	0.25	55.3
	0.50	98.9
	0.75	99.1
Undaria pinnatifida (Wakame)	0.25	21.0
	0.50	45.8
	0.75	61.6
Hizikia fusiformis (Hijiki)	0.25	41.3
	0.50	57.8
	0.75	85.4

Table 4.23. *Effect of soluble brown algae fibers on pepsin inhibition (Horie et al. 1995)*

Regardless of the nature or origin of the fibers, a dose effect is also observed in pepsin inhibition. Soluble brown algae fibers also have a significantly greater pepsin inhibitory effect than that reported for mushroom fibers consumed in Asia, such as Shiitake, Hiratake and Yanagimatsutake (Horie et al. 1995). As an indication, soluble fibers from Hiratake inhibit 87% of pepsin activity when added at a level of 0.75% of the mixture. At the same concentration, soluble fibers from Kombu inhibit 99% of pepsin activity.

The presence of factors inhibiting the enzymatic activity of digestive proteases, and even human intestinal juices, has been demonstrated in the red algae *Palmaria palmata* (Dulse) (Galland-Irmouli et al. 1999).

Trypsin and chymotrypsin activities are affected in the presence of an algal extract (–30% BSA hydrolysis activity). The most remarkable result is obtained in intestinal juices, where the presence of *P. palmata* extract inhibits nearly 45% of their protease activity (see Figure 4.29).

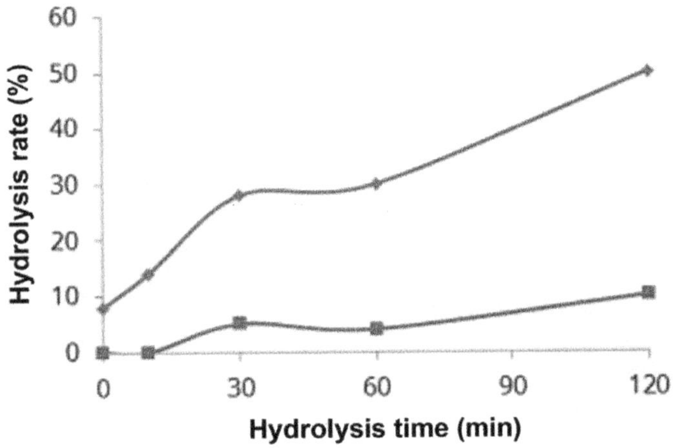

Figure 4.29. *In vitro digestibility of BSA (blue curve) and BSA +* Palmaria palmata *extract (red curve) (Galland-Irmouli et al. 1999; Fleurence 2018)*

4.7.2. *Proteins*

The nutritional value of proteins depends on their amino acid composition and, above all, on the balance between essential and non-essential amino acids. It also depends on the digestibility of these proteins, i.e. their ability to be broken down by digestive proteases and therefore metabolized.

The in vitro digestibility of the protein fraction of many edible algae has been studied. Digestion by pepsin, a protease found in gastric juice, appears to be low, or even moderate, for most species (Table 4.23). The red alga *P. tenera* (Nori) appears to be the most easily digested by pepsin and pancreatin. This gives the protein fraction of this alga a higher nutritional value, given its greater sensitivity to digestive enzymes, pronase being a protease of fungal origin.

Species	Digestion percentage by pepsin	Digestion percentage by pancreatin	Digestion percentage by pronase
Ulva pertusa	17.0	66.6	94.8
Codium fragile	22.5	37.2	70.9
Laminaria japonica	39.0	54.0	83.9
Undaria pinnatifida	23.9	48.1	87.2
Porphyra tenera	56.7	56.1	78.4
Palmaria palmata	–	56.0	–
Grateloupia turuturu	15.8	34.0	59.2

Table 4.24. *In vitro digestibility of the protein fraction of some edible algae (adapted from Fujiwara-Arasaki et al. (1984) and Galland-Irmouli et al. (1999))*

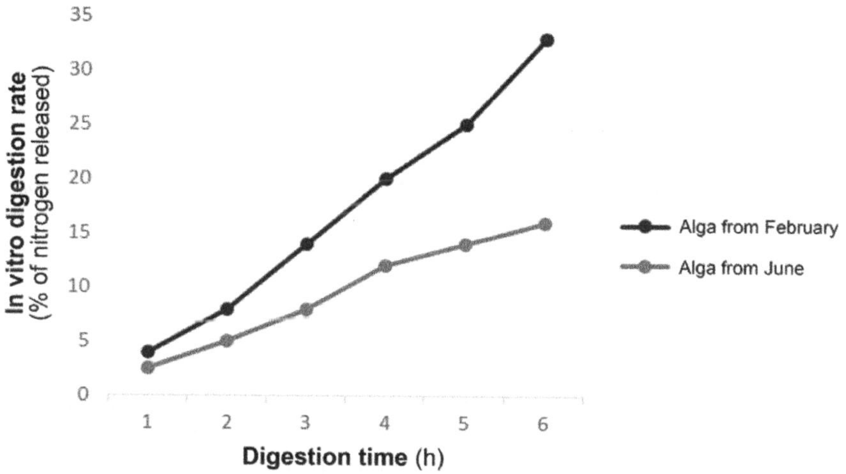

Figure 4.30. *Influence of season on the digestion of* Palmaria palmata *(% nitrogen released) (Galland-Irmouli et al. 1999)*

In the alga *Palmaria palmata* (Dulse), the in vitro digestibility of the protein fraction differs according to the season. It is notably higher at the end of winter (February) than at the beginning of summer (June) (see Figure 4.30). Casein, a reference protein used to study in vitro protein digestibility, is less sensitive to proteolysis when an extract of *P. palmata* is added to the reaction medium containing pepsin and pancreatin. This decrease in the casein digestion activity is particularly noticeable after six hours of digestion when 25% of algal extract (weight/volume) is added (see Figure 4.31).

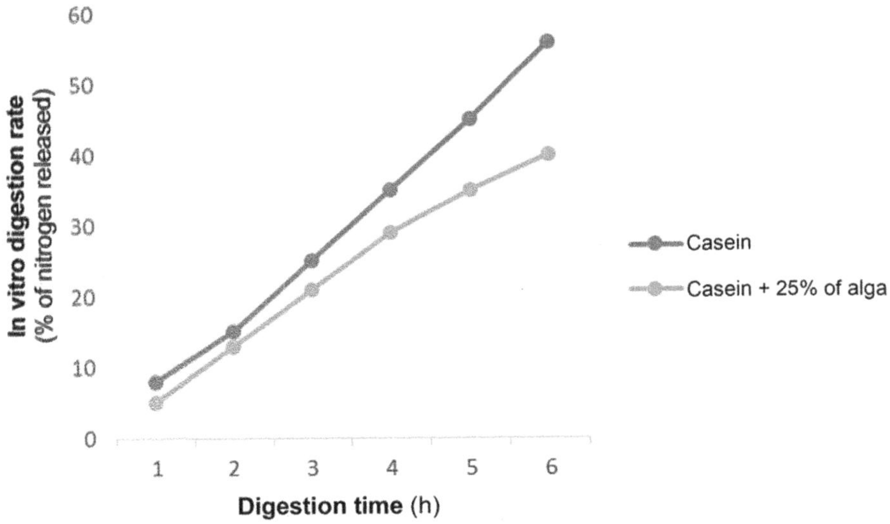

Figure 4.31. *Influence of the addition of* Palmaria palmata *extract on the in vitro digestion of casein (% nitrogen released) (Galland-Irmouli et al. 1999)*

XYLANS AND CELLULOSE, POLYSACCHARIDES FOUND IN THE ALGA *PALMARIA PALMATA*. –

The polysaccharides in *Palmaria palmata*, xylans and cellulose, appear to limit protein digestibility. They are therefore considered antinutritional factors for the assimilation of this alga's proteins by the human body.

However, this finding must be weighed up, as the inhibition of protein digestion by algal fibers depends heavily on the biochemical nature of the polysaccharides or fibers in question. Indeed, the in vitro digestibility over six hours of proteins from *Gracilaria verrucosa* (Ogonori) is significantly higher (approximately three times) than that recorded for proteins from the alga *Palmaria palmata* (Dulse) (see Figure 4.32).

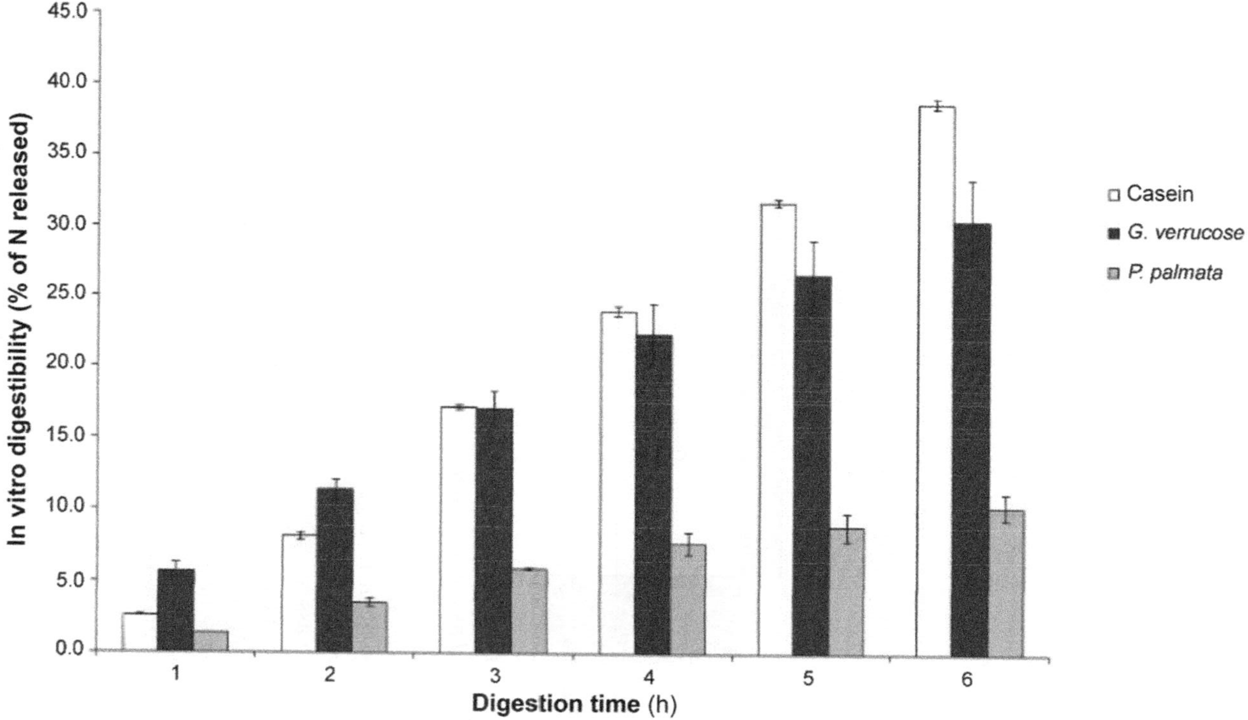

Figure 4.32. *Comparison of the in vitro digestibility of the protein fraction of two edible red algae (*Gracilaria verrucosa, Palmaria palmata*) and casein (Marrion et al. 2005)*

This apparent difference in digestibility is due to the different biochemical nature of the fibers and their content in each species (Table 4.25).

The species *G. verrucosa* is characterized by a total fiber content representing 31.3% of the dry algal matter. In contrast, the total fiber content reaches 56.5% in *P. palmata*, a species for which the protein fraction is less easily digested in vitro (pepsin + pancreatin).

The difference between the two species also concerns the nature of the fibers and their respective distributions. *G. verrucosa* has very few insoluble fibers (0.8%) and a soluble fiber content that is one-third lower than that reported for *P. palmata*.

The biochemical nature of the fibers also differs between the two algae. The fibers of *G. verrucosa* are mainly composed of hexoses and few pentoses. This is not the case for the fibers of *P. palmata*, which are mainly composed of pentoses (Table 4.25).

	Gracilaria verrucosa	*Palmaria palmata*
Total fiber	31.3%	56.5%
Soluble fiber	305%	45.3%
Insoluble fiber	0.8%	11.2%
Hexoses	30.5%	5.3%
Pentoses	–	39.0%

Table 4.25. *Nature and fiber content of* G. verrucosa *and* P. palmata *algae (expressed as a percentage of dry matter) (Marrion et al. 2005)*

The nature and content of algal fibers therefore play an essential role in the digestibility and feed availability of the algal protein fraction. These fibers are determining factors in assessing the nutritional quality of algal proteins and, beyond that, other nutrients such as minerals. In addition, some authors recommend evaluating the indigestible fraction (total fiber) of algae in order to obtain an initial assessment of their nutritional quality (Rupérez and Toledano 2003).

4.7.3. *Minerals*

Algae are a significant source of minerals, particularly microelements and even trace elements. The mineral content of these organisms varies greatly depending on the season, species and processing methods (drying, rehydration, cooking, freezing-thawing, canning, fermentation, etc.). In this context, the mineral nutritional quality of algae can only be assessed on a case-by-case basis, i.e. product by product. The specific case of iodine fully illustrates this point.

Iodine is an important mineral from a nutritional point of view. The minimum recommended daily intake is 150 µg of iodine. However, the maximum threshold is the subject of much debate. It is normally set at 600 µg per day (Chapter 6, section 6.6). Algae are rich in iodine, particularly brown alga (Table 4.26). These levels vary greatly depending on the species and the season of harvest (Table 4.26).

Species	Total iodine content (µg/g DM)
Fucus vesiculosus	226–732
Sargassum fusiforme (Hijiki)	58–629
Himanthalia elongata (Sea wrack)	59–605
Laminaria digitata (Breton Kombu)	1,997–10,000
Laminaria (Saccharina) japonica (Kombu)	241–4,300
Laminaria saccharina (Saccharina latissima) (Royal Kombu)	238–6,568
Laminaria longicruris (Atlantic Kombu)	764–1,879
Alaria esculenta (Irish Wakame, Atlantic Wakame)	95–1,844
Undaria pinnatifida (Wakame)	31–514
Eisenia bicyclis (Arame)	586–600

Table 4.26. *Examples of iodine content in some brown algae species frequently used in human food (Blikra et al. 2022)*

The iodine content of brown algae when harvested appears to be a barrier to their direct consumption. The toxicity threshold for iodine is estimated at a daily intake of 1,000 µg/day. However, this value is not accepted by many toxicologists, who consider it excessive. Compliance with this threshold theoretically limits the consumption of certain algae to 1 or 1.5 g of dry product (5 g of wet product), or even less for species that have been consumed for thousands of years, such as Kombu (Table 4.26).

However, the algae available to consumers generally undergo a processing procedure that involves dehydration through drying for varying lengths of time, followed by rehydration and cooking before consumption. The rehydration and cooking stages significantly reduce the iodine content of algae that undergo this process. In the brown alga *Alaria esculenta*, this process removes 62% of the iodine through a leaching phenomenon. It is also effective for the deiodination of the red alga *Palmaria palmata* and the green alga *Ulva intestinalis*. It produces a 15% reduction in iodine for the former and a 10% reduction for the latter (see Figure 4.33) (Nitschke and Stengel 2016). Certain operations, such as blanching algae, are particularly effective in removing iodine from Laminaria (Chapter 6, section 6.6).

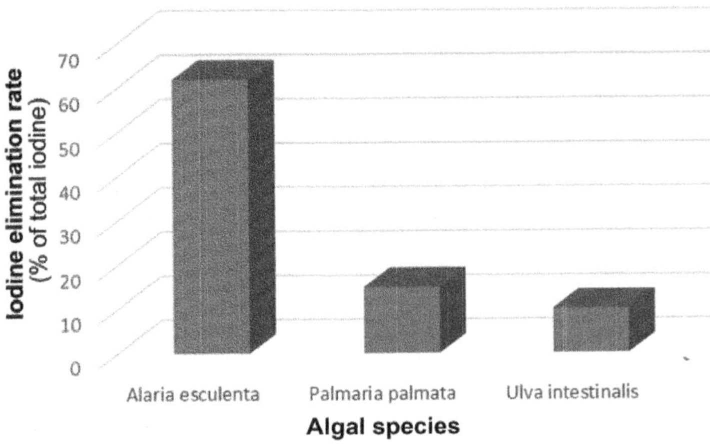

Figure 4.33. *Impact of dehydration and rehydration in ionized water on the iodine removal rate of three edible algae,* Alaria esculenta *(Atlantic Wakame),* Palmaria palmata *(Dulse) and* Ulva intestinalis *(Aonori) (Nitschke and Stengel 2016)*

The iodine content of processed products is not the only criterion for assessing the benefits or risks of using algae as a source of iodine in the diet. Evaluating the bioavailability of this element and, above all, its bioaccessibility is crucial. The intestinal absorption or bioaccessibility of iodine differs depending on the algal species and therefore on the chemical forms of this element (Dominguez-Gonzalez et al. 2017) (Chapter 6, section 6.6).

4.7.4. *In vivo studies*

Nutritional quality, as previously described, takes into account the biochemical composition, characteristics and in vitro digestibility of certain compounds. The in vivo bioavailability and bioaccessibility of nutrients are the only relevant criteria for assessing the nutritional value of a product, and therefore of algae. The evaluation of bioavailability requires the implementation of in vivo studies. This type of study is mainly carried out on murine models.

In vivo studies conducted on murine models concern the bioavailability of mineral, protein or lipid nutrients in the presence of algae in the diet. The nutrients involved in this type of study are not always of algal origin, but are provided by other sources used in the diets tested. These studies shed direct light on the action of algal antinutritional factors, fiber and the assimilation of compounds of nutritional interest.

A few studies on the bioavailability of nutrients in vivo are available. These studies, which are few in number, have been conducted on rats and often concern mineral bioavailability and sometimes the assimilation of other nutrients such as proteins or lipids.

4.7.4.1. *The bioavailability of minerals in the presence of algae in the diet*

Minerals, and more particularly trace elements, are constituents that contribute significantly to the nutritional quality of a diet. In the presence of algae, their bioavailability depends mainly on the ability of algal fibers to bind them. This latter point is far from being fully understood and many uncertainties remain depending on the minerals and, above all, their chemical speciation.

However, a fairly comprehensive study has been carried out on two species of edible algae that are frequently consumed worldwide: the red alga *Porphyra tenera* (Nori) and the brown alga *Undaria pinnatifida* (Wakame) (Urbano and Goni 2002). This study evaluates the in vivo bioavailability of nutrients (proteins, minerals) in the presence of these algae. It was conducted on rats that were fed a diet containing algae or cellulose as a source of fiber (control diet) for two weeks.

Animals fed an alga-based diet developed lower apparent digestibility and retention capacity for magnesium and iron than animals fed the control diet. These elements were found in higher quantities in the feces and urine of animals fed algae (Table 4.27).

	Control diet	"Nori" diet	"Wakame" diet
Magnesium intake from food (mg)	270	175	219
Fecal excretion of magnesium (mg)	46	62	113
Urinary excretion of magnesium (mg)	19	17	25
Apparent retention (%)	75.90	54.70	35.80

Table 4.27. *Influence of the addition of algae (Nori or Wakame) in the diet of rats on the in vivo assimilation of magnesium (Urbano and Goni 2002)*

However, there were no significant differences in calcium assimilation between animals fed different diets. This last point clearly shows that the chemical speciation of minerals also affects their ability to be bound by fibers and therefore to be more or less available for assimilation by the body.

The non-assimilation process is also more significant with Wakame than with Nori. This finding highlights the importance of the origin and biochemical nature of fibers in their effectiveness as antinutritional factors.

4.7.4.2. *The bioavailability of proteins in the presence of algae in the diet*

As previously indicated, algal fibers have the ability to inhibit the action of certain digestive enzymes (pepsin, pancreatin, human intestinal juices) in vitro. This property is responsible for the in vitro limitation of the digestibility of reference dietary proteins such as casein, as well as for the low in vitro digestion of algal proteins.

Studies conducted on rats show that adding algae (Nori or Wakame) at a level of 147 $g.kg^{-1}$ of experimental diet has a significant effect on the fecal excretion of ingested proteins (Table 4.28). A 6.5% loss in the digestibility of the ingested protein fraction was observed in rats fed a diet supplemented with algae. These in vivo results confirm the negative influence of algal fiber on digestion and therefore on protein assimilation, regardless of the origin of the protein.

	Control diet	"Nori" diet	"Wakame" diet
Protein intake (g/week)	19.3	19.7	21.7
Fecal protein excretion (g/week)	1.42	2.73	2.94
Urinary excretion (g/week)	6.49	5.59	4.11
Apparent digestibility (%)	92.60	86.10	86.20

Table 4.28. *Influence of the addition of algae (Nori or Wakame) in the diet of rats on the in vivo assimilation of proteins (according to Urbano and Goni 2002)*

4.7.4.3. *The bioavailability of lipids with the presence of algae in the diet*

Very little research has been conducted on the in vivo assimilation of lipids and their derivatives. The addition of algae to the diet of rats appears to have an effect on fecal excretion of lipids and their apparent digestibility. The latter is lower when rats are fed a diet containing Nori or Wakame (Urbano and Goni 2002) (Table 4.29).

However, the difference observed between the two types of diet (without algae and with algae) remains minor (approximately 1%). This suggests that the presence of algae fiber has a very limited effect on the in vivo assimilation of the lipid fraction. This finding contrasts sharply with that established for proteins.

	Control diet	"Nori" diet	"Wakame" diet
Lipid intake (g/week)	5.45	5.48	6.08
Lipid excretion (g/week)	0.05	0.10	0.11
Apparent digestibility (%)	99.00	98.10	98.14

Table 4.29. *Influence of adding algae in the diet of rats on the in vivo assimilation of lipids (according to Urbano and Goni 2002)*

The nutritional quality of algae depends on their biochemical and mineral composition, as well as on the bioavailability of the nutrients they contain. Most published studies focus on in vitro models. For in vivo models, studies mainly focus on the impact of algae consumption on the diet (sections 4.7.4.1–4.7.4.3).

Polysaccharides, major compounds in the biochemical composition of algae, behave like dietary fiber. They act as antinutritional factors, limiting the in vitro digestibility of the algal protein fraction, as well as any other protein fraction provided by diet.

They can therefore limit the nutritional value of algae as a source of dietary protein. However, this finding must be qualified, as the inhibitory effect of algal fibers on protein digestion depends heavily on the origin and biochemical nature of these fibers (see Figure 4.32 and Table 4.25).

Furthermore, enzymatic treatment of these fibers can also significantly improve the digestibility of the algal protein fraction (Chapter 5, section 5.8). In the same context, the bioavailability of minerals is affected differently depending on the nature of the fibers and their chemical speciation. Lipids, on the other hand, appear to be less sensitive to the presence of algal fibers that limit their assimilation.

The nutritional quality of algae therefore depends mainly on the fibers and their ability to limit, to a greater or lesser extent, the assimilation of other nutrients of interest (proteins, lipids, vitamins, minerals and carotenoids). The effect of fiber on vitamin assimilation is largely unknown, but it can be assumed to be minor, particularly with regard to fat-soluble vitamins (vitamins A, D and E). For water-soluble vitamins (vitamins C and B), the question remains to be addressed.

Finally, the high content of soluble fiber in algae makes them an original and complementary source to other plant sources, particularly terrestrial ones. Fiber remains an important component in nutrition, facilitating intestinal transit, among other things, and as such, soluble fiber from algae also contributes to the nutritional quality of edible algae.

4.8. References

Ale, M.T. and Meyer, A.S. (2013). Fucoidans from brown seaweeds: An update on structures, extraction techniques and use of enzymes as tools for structural elucidation. *Royal Society of Chemistry Advances*, 3, 8131–8141.

Amorin-Carrilho, K., Lage-Yusty, M.A., Lopez-Hernandez, J. (2014). Variation of bioactive compounds in dried seaweed *Himanthalia elongata* subjected to different culinary processes. *CyTA Journal of Food*, 12, 336–339.

Arasaki, S. and Arasaki, T. (1985). *Les légumes de mer. Comment être et paraître en forme.* Éditions de la Maisnie, Douarnenez.

Bandarra, N.M., Batista, I., Nunes, M.L., Empis, J.M., Christie, W.W. (1997). Seasonal changes in lipid composition of sardine (*Sardina pilchardus*). *Journal of Food Science*, 62(1), 40–42.

Blikra, M.J., Henjum, S., Aakre, I. (2022). Iodine from brown algae in human nutrition, with an emphasis on bioaccessibility, bioavailability, chemistry, and effects of processing: A systematic review. *Comprehensive Reviews in Food Science and Food Safety*, 21(2), 1517–1536.

Boulom, S., Robertson, J., Hamid, N., Ma, Q., Lu, J. (2014). Seasonal changes in lipid, fatty acid, α-tocopherol and phytosterol contents of seaweed, *Undaria pinnatifida*, in the Marlborough Sounds, New Zealand. *Food Chemistry*, 161, 261–269.

Bruni, G. and Stancher, B. (1973). Valutazione della qualita della frazione proteica di alcune alghe dell'alto Adriatico. *Rassegna Chimica*, 3, 225–231.

Darcy-Vrillon, B. (1993). Nutritional aspects of the developing use of marine macroalgae for the human food industry. *International Journal of Food Sciences and Nutrition*, 1993, 44(1), S23-S35 ref. 47.

Dembitsky, V., Rozentsvet, O.A., Penchenkina, E.E. (1990). Glycolipids, phospholipids and fatty acids of brown algae species. *Phytochemistry*, 11, 3417–3421.

Deniaud, E. (2002). Structure et interaction des polymères dans les parois cellulaires de *Palmaria palmata* (Rhodophyta). PhD Thesis, Université Paris VI.

Denis, C., Morançais, M., Li, M., Deniaud, E., Gaudin, P., Wielgosz-Collin, G., Barnathan, G., Jaouen, P., Fleurence, J. (2010). Study of the chemical composition of edible red macroalgae *Grateloupia turuturu* from Brittany (France). *Food Chemistry*, 119, 913–917.

Domínguez-González, M.R., Chiocchetti, G.M., Herbello-Hermelo, P., Velez, D., Devesa, V., Bermejo-Barrera, P. (2017). Evaluation of iodine bioavailability in seaweed using in vitro methods. *Journal of Agricultural and Food Chemistry*, 65(38), 8435–8442.

Dumay, J. and Morançais, M. (2016). Proteins and pigments. In *Seaweed in Health and Disease Prevention*, Fleurence, J. and Levine, I. (eds). Academic Press, Elsevier, Amsterdam.

Fleurence, J. (1999a). Seaweed proteins: Biochemical, nutritional aspects and potential uses. *Trends in Food Science and Technology*, 10, 25–28.

Fleurence, J. (1999b). The enzymatic degradation of algal cell walls: A useful approach for improving protein accessibility? *Journal of Applied Phycology*, 11, 313–314.

Fleurence, J. (2004). Seaweed proteins. In *Proteins in Food Processing*, Yada, R.Y. (ed.). Woodhead Publishing Limited, Cambridge.

Fleurence, J. (2018). *Les algues alimentaires : bilan et perspectives*. Lavoisier Tec & Doc, Paris.

Fleurence, J. and Le Cœur, C. (1994). Influence of digestion procedures on the determination of Lead and Cadmium levels in the laminariale *Undaria pinnatifida* (Wakame) by flame atomic absorption spectrophotometry. *Botanica Marina*, 37, 555–559.

Fleurence, J., Gutbier, G., Mabeau, S., Leray, C. (1994). Fatty acids from 11 marine macroalgae of the French Brittany coast. *Journal of Applied Phycology*, 6, 527–532.

Fleurence, J., Le Cœur, C., Mabeau, S., Maurice, M., Landrein, A. (1995). Comparison of different extractive procedures for proteins from the edible seaweeds *Ulva rigida* and *Ulva rotundata*. *Journal of Applied Phycology*, 7, 577–582.

Fleurence, J., Chenard, E., Luçon, M. (1999). Determination of the nutritional value of proteins obtained from *Ulva armoricana*. *Journal of Applied Phycology*, 11, 231–239.

Fleurence, J., Morançais, M., Dumay, J., Decottignies, P., Turpin, V., Munier, M., Garcia-Bueno, N., Jaouen, P. (2012). What are the prospects for using seaweed in human nutrition and for marine animals raised through aquaculture? *Trends in Food Science and Technology*, 27, 57–61.

Fleurence, J., Morançais, M., Dumay, J. (2018). Seaweed proteins. In *Proteins in Food Processing*, Yada, R.Y. (ed.). Woodhead Publishing, Cambridge.

Francavilla, M., Manara, P., Kamaterou, P., Monteleone, M., Zabaniotou, A. (2015). Cascade approach of red macraoalgae *Gracilaria gracilis* sustainable valorization by extraction of phycobiliproteins and pyrolysis of residue. *Bioresource Technology*, 184, 305–313.

Fujiwara-Arasaki, T., Mino, N., Kuroda, M. (1984). The protein value in human nutrition of edible marine algae in Japan. *Hydrobiologia*, 116(117), 513–516.

Galland-Irmouli, A.V., Fleurence, J., Lamghari, R., Luçon, M., Rouxel, C., Barbaroux, O., Bronowicki, J.P., Villaume, C., Guéant, J.L. (1999). Nutritional value of proteins from *Palmaria palmata* (Dulse). *Journal of Nutritional Biochemistry*, 10, 353–359.

Galland-Irmouli, A.V., Pons, L., Luçon, M., Villaume, C., Mrabet, J., Guéant, J.L., Fleurence, J. (2000). One step purification from the red macroalga *Palmaria palmata* using preparative polycarylamide gel electrophoresis. *Journal of Chromatography B*, 739, 117–123.

Guizani, S.E.O. and Moujahed, N. (2015). Seasonal variation of chemical and fatty acids composition in Atlantic mackerel from the Tunisian northern-east coast. *Journal of Food Process Technology*, 6(9), 1–4.

Haug, A., Larsen, B., Smisrod, O. (1974). Uronic acid sequence in alginate from different sources. *Carbohydrates Research*, 32, 217–225.

Hong, D.D., Hien, H.M., Son, P.N. (2007). Seaweeds from Vietnam used for functional food, medicine and biofertilizer. *Journal of Applied Phycology*, 19, 817–826.

Horie, Y., Sugase, K., Horie, K. (1995). Physiological differences of soluble and insoluble dietary fibre fractions of brown algae and mushrooms in pepsin activity and protein digestibility. *Asia Pacific Journal of Clinical Nutrition*, 4, 251–255.

Hotimchenko, S.V. (2002). Fatty acid composition of algae from habitats with varying amounts of illumination. *Russian Journal of Marine Biology*, 28, 218–220.

Idler, D.R. and Wiseman, P. (1970). Sterols in red algae (Rhodophyceae): Variation in demosterol content of Dulse (*Rhodymenia palmata*). *Comparative Biochemistry Physiology*, 35, 679–687.

Jurkovic, N., Kolb, N., Colic, I. (1995). Nutritive value of marine algae *Laminaria japonica* and *Undaria pinnatifida*. *Die Nahrung*, 39, 63–66.

Kendel, M., Couzinet-Mossion, A., Viau, M., Fleurence, J., Barnathan, G., Wielgosz-Collin, G. (2013). Seasonal composition of lipids, fatty acids and sterols in the edible red alga *Grateloupia turuturu*. *Journal of Applied Phycology*, 25, 425–432.

Kendel, M., Wielgosz-Collin, G., Bertrand, S., Roussakis, C., Bourgougnon, N., Bedoux, G. (2015). Lipid composition, fatty acids and sterols in the seaweeds *Ulva armoricana* and *Soliera chordalis* from Brittany (France): An analysis from nutritional, chemotaxonomic and antiproliferative activity perspectives. *Marine Drugs*, 1, 5606–5628.

Khairy, H.N. and El-Shafay, M. (2013). Seasonal variations in the biochemical composition of some common seaweed species from the coast of Abu Qir Bay, Alexandria, Egypt. *Oceanologia*, 55, 435–452.

Khotimchenko, S.V. and Vaskovsky, V.E. (1990). Distribution of polyenoic fatty acids in red macrophytic algae. *Botanica Marina*, 33, 525–528.

Kloareg, B. and Quatrano, R.S. (1988). Structure of the cell walls of marine algae and ecophysiological functions of the matrix polysaccharides. *Oceanographic Annual Review of Marine Sciences*, 26, 259–315.

Komalavalli, N. and Latitha, N. (2015). Proximate composition and amino acid profile of five green algal seaweeds from Mandapam coastal regions, Tamil Nadu, India. *International Journal of Advances in Interdisciplinary Research*, 2, 37–40.

Lahaye, M. (1991). Marine algae as sources of fibres: Determination of soluble and insoluble dietary fibre contents in some "sea vegetables". *Journal of Food Science and Agriculture*, 54, 587–594.

Larsen, B., Haug, A., Painter, T. (1966). Sulphated polysaccharides in brown algae. I. Isolation and preliminary characterization of three sulphated polysaccharides from *Ascophyllum nodosum* (L.) Le Jol. *Acta Chemica Scandinavica*, 20, 219–230.

Larsen, B., Haug, A., Painter, T. (1970). Sulfated polysaccharides in brown algae. *Acta Chemica Scandinavica*, 24, 3339–3352.

Lechat, H., Amat, M., Mazoyer, J., Buléon, A., Lahaye, M. (2000). Structure and distribution of glucomannan and sulfated glucan in the cell walls of the red alga *Kappaphycus alvarezii* (Gigartinales, Rhodophyta). *Journal of Phycology*, 36, 891–902.

Mabeau, S. and Fleurence, J. (1993). Seaweeds in food products: Biochemical and nutritional aspects. *Trends in Food Sciences and Technology*, 4, 103–107.

MacArtain, P., Gill, C.I., Brooks, M., Campbell, R., Rowland, I.R. (2007). Nutritional value of edible seaweeds. *Nutrition Reviews*, 65, 535–543.

MacPherson, M.G. and Young, E.G. (1949). The chemical composition of marine algae. *Canadian Journal Research*, 27, 73–77.

Marrion, O., Fleurence, J., Schwertz, A., Guéant, J.L., Mamelouk, L., Ksouri, J., Villaume, C. (2005). Evaluation of protein in vitro digestibility of *Palmaria palmata* and *Gracilaria verrucosa*. *Journal of Applied Phycology*, 17, 99–102.

Matanjun, P., Mohamed, S., Mustapha, N.M., Muhammad, K. (2009). Nutrient content of tropical edible seaweeds, *Eucheuma cottonii*, *Caulerpa lentillifera* and *Sargassum polycystum*. *Journal of Applied Phycology*, 21, 75–80.

Medcalf, D. and Larsen, B. (1977a). Fucose-containing polysaccharides in the brown algae *Ascophyllum nodosum* and *Fucus vesiculosus*. *Carbohydrate Research*, 59, 531–537.

Medcalf, D. and Larsen, B. (1977b). Structural studies on Ascophyllan and the fucose-containing complexes from the brown alga *Ascophyllum nodosum*. *Carbohydrate Research*, 59, 539–546.

Mizuno, H., Saito, T., Iso, N., Onda, N., Noda, K., Takada, K. (1983). Mannuronic to Guluronic acid ratios of Alginic acids prepared from various brown seaweeds. *Bulletin of the Japonese Society of Scientific Fisheries*, 49, 1591–1593.

Morgan, K.C., Wright, J.L.C., Simpson, F.J. (1980). Review of chemical constituents of the red alga *Palmaria palmata* (Dulse). *Economic Botany*, 34, 27–50.

Morrice, L.M., McLean, M., Long, W., Williamson, F.B. (1983). Porphyran primary structure. An investigation using B-agarase I from *Pseudomonas atlantica* and 13 C-NMR Spectroscopy. *European Journal of Biochemistry*, 133, 673–684.

Munda, I. (1972). On the chemical composition, distribution and ecology of some common benthic marine algae from Iceland. *Botanica Marina*, 15, 1–45.

Newton, I.S. (1996). Long chain fatty acids in health and nutrition. *Journal of Food Lipids*, 3, 233–249.

Nisizawa, K., Noda, H., Kikuchi, R., Watanabe, T. (1987). The main seaweed foods in Japan. *Hydrobiologia*, 151(152), 5–29.

Nitschke, U. and Stengel, D.B. (2016). Quantification of iodine loss in edible Irish seaweeds during processing. *Journal of Applied Phycology*, 28(6), 3527–3533.

Norziah, M.H. and Ching, C.Y. (2000). Nutritional composition of edible seaweed *Gracilaria changii. Food Chemistry*, 68, 69–76.

O'Leary, F. and Samman, S.S. (2010). Vitamin B12 in health and disease. *Nutrients*, 2, 229–316.

Patarra, R.F., Paiva, L., Neto, A.L., Lima, E., Baptista, J. (2011). Nutritional value of selected macroalgae. *Journal of Applied Phycology*, 23, 205–208.

Peñalver, R., Lorenzo, J.M., Ros, G., Amarowicz, R., Pateiro, M., Nieto, G. (2020). Seaweeds as a functional ingredient for a healthy diet. *Marine Drugs*, 18(6), 301. doi: 10.3390/md18060301.

Percival, E. (1979). The polysaccharides of green, red and brown seaweeds: Their basic structure, biosynthesis and function. *British Phycological Society*, 14, 103–107.

Perez, R. (1997). *Ces algues qui nous entourent.* Ifremer, Paris.

Ratana-Arpon, P. and Chirapart, A. (2006). Nutritional evaluation of tropical green seaweeds *Caulerpa lentillefera* and *Ulva reticula*. *Kasetsart Journal*, 40, 75–83.

Rodrigues, D., Freitas, A.C., Pereira, L., Rocha-Santos, T.A.P., Vasconcelos, M.W., Roriz, M., Rodriguez-Alcala, L.M., Gomes, A.M.P., Duarte, A.C. (2015). Chemical composition of red, brown and green macroalgae from Buarcos bay in Central West Coast of Portugal. *Food Chemistry*, 183, 197–207.

Rupérez, P. and Toledano, G. (2003). Indigestible fraction of edible marine seaweeds. *Journal of the Science of Food and Agriculture*, 83, 1267–1272.

Schmid, M., Guihéneuf, F., Stengel, D. (2014). Fatty acid contents and profiles of 16 macroalgae collected from the Irish Coast at two seasons. *Journal of Applied Phycology*, 26, 451–463.

Stiger-Pouvreau, V., Bourgougnon, N., Deslandes, E. (2016). Carbohydrates from seaweeds. In *Seaweed in Health and Disease Prevention*, Fleurence, J. and Levine, I. (eds). Academic Press, Elsevier, Amsterdam.

Tanaka, Y., Ikebe, K., Tanaka, R., Kunita, N. (1973). Contents of heavy metals in foods. *Shoku Eishi*, 14, 196–201.

Urbano, M.G. and Goni, I. (2002). Bioavailability of nutrients in rats fed on edible seaweeds Nori (*Porphyra tenera*) and Wakame (*Undaria pinnatifida*) as a source of dietary fibre. *Food Chemistry*, 76, 281–286.

Vera, J., Castro, J., Gonzalez, A., Moenne, A. (2011). Seaweed polysaccharides and derived oligosaccharides stimulate defense response and protection against pathogens in plants. *Marine Drugs*, 9, 2514–2522.

Viekbe, C., Borgström, J., Piculell, L. (1995). Characterisation of kappa- and iota-carrageenan coils and helices by MALLS/GPC. *Carbohydrate Polymers*, 27, 145–154.

Wang, H., Zhu, B., Yao, Z. (2023). Porphyran, porphyran oligosaccharides and porphyranase: Source, structure, preparation methods and applications. *Algal Research*, 73, 103167. doi: 10.1016/j.algal.2023.103167.

Watanabe, F., Takenaka, S., Katsura, H., Masumder, S.A.M.Z.H., Abe, K., Tamura, Y., Nakano, Y. (1999). Dried green and purple lavers (Nori) contain substantial amounts of biological active vitamin B12 but less dietary iodine relative to other edible seaweeds. *Journal of Agricultural Food Chemistry*, 47, 2341–2343.

Watanabe, F., Takenaka, S., Katsura, H, Miyamoto, E., Abe, K. Tamura, Y., Nakatsuka, T., Nakano, Y. (2000). Characterization of a vitamin B12 compound in the edible purple laver *Porphyra yezoensis*. *Biosciences, Biotechnology, Biochemistry*, 64, 2712–2715.

Wielgosz-Collin, G., Kendel, M., Couzinet-Mossion, A. (2016). Lipids, fatty acids, glycolpids and phospholipids. In *Seaweed in Health and Disease Prevention*, Fleurence, J. and Levine, I. (eds). Academic Press, Elsevier, Amsterdam.

Yeh, T.S., Hung, N.H., Lin, T.C (2014). Analysis of iodine content in seaweed by GC-ECD and estimation of iodine intake. *Journal of Food and Drug Analysis*, 22, 189–196.

5

Processing Methods

5.1. Introduction

Algae processing methods are based on unit treatment operations that generally involve an initial washing stage (in salt water or fresh water) followed by a raw material stabilization stage.

The raw material is preserved through simple operations such as drying, salting or freezing. These operations are considered algae processing methods. They respond to an industrial reality but are not viewed in the same way from a regulatory standpoint. European and French regulations in particular do not consider freezing to be a processing activity.

The products resulting from this initial processing can then undergo more elaborate processes such as appertization, fermentation, maturation, enzymatic liquefaction or liquid extraction. This second type of processing is often carried out in Western countries where the culinary use of dried algae is relatively undeveloped.

5.2. Drying

Drying is a process used to transform and preserve algae. It is generally carried out on algae that have been washed in either sea water or fresh water. The purpose of washing is to remove epiphytes, small benthic fauna, sand and salt from the algae to be processed.

Algae can be dried in various ways, but there are two main methods:

– a gentle method based on drying the algae in the sun and in the open air (a traditional method originating in Asia);

For a color version of all the figures in this chapter, see www.iste.co.uk/fleurence/marinealgae.zip

Marine Algae, A Sustainable Food Resource,

– a method based on moderate heating (30–45°C) in enclosed spaces or "tobacco dryers".

5.2.1. *Sun drying or the traditional method*

This method is mainly used in Asia. It is largely used in China and Japan for drying *Laminaria japonica* (*Saccharina japonica*) (Kombu). The algae are spread out on the shore and dried in the sun (see Figure 5.1). The plants are turned every four hours to ensure even dehydration. In the evening, the algae are covered with straw mats to prevent nighttime humidity and morning dew from rehydrating the plant.

Figure 5.1. *Sun drying of* Laminaria japonica *(*Saccharina japonica*) (photo credit © Barbaroux O. 2016)*

Exposure to sunlight is prolonged until the water content is close to 15% of the algal mass (Perez 1997). If weather conditions are unfavorable, sun drying is postponed. In this case, the algae undergo a preservative treatment. The blades are boiled for 20 seconds, and then soaked in brine before being drained. The algae treated in this way are stacked in layers in a tray, and coarse salt is placed between the layers to prevent the action of bacterial flora. The algae treated in this way are then subjected to traditional drying as soon as the sun returns.

Sun drying is a traditional method used in Japan to produce different varieties of Kombu for human consumption (Table 5.1). The species concerned are not limited to *Laminaria japonica* (*Saccharina japonica*). Other laminaria species are also involved, and the products obtained have trade names that are well known to consumers.

Algae species	Trade name
Kjellmaniella crassifolia	Gagome-Kombu
Laminaria angustata var. *longissima*	Naga-Kombu
Laminaria religiosa	Hosome-Kombu
Laminaria japonica (*Saccharina japonica*)	Ma-Kombu
Laminaria angustata	Mitsuishi-Kombu
Laminaria japonica var. *ochotensis*	Rishiri-Kombu

Table 5.1. *Different trade names for sun-dried Kombu-type products (Nisizawa et al. 1987)*

Sun drying may be the first step in a more complex process leading to a wide variety of processed products with highly codified trade names (Table 5.2).

Type of product	Trade name
Shredded Kombu	Kizami-Kombu
Sliced Kombu	Oboro-Kombu
Salted Kombu	Shio-Kombu
Salted and boiled Kombu	Tsukudani-Kombu
Kombu roll	Kombu-maki
Powdered Kombu	Kombu-cha

Table 5.2. *Different trade names for processed products obtained from sun-dried Kombu (Nisizawa et al. 1987)*

The brown alga *Hizikia fusiformis*, traditionally consumed in Japan under the name Hiziki (Hijiki), can also be sun-dried. When subjected to this process, it is listed under the name "Sobushi-Hiziki". As before, drying is often the first step in a processing method. In the case of Hiziki, the dried algae can be boiled in a copper cauldron for several hours with *Eisenia* fronds. The product resulting from this process is called "Hoshi-Hiziki" (Nisizawa et al. 1987). Green algae frequently consumed in Japan are often dried in the sun. They are marketed under the generic name "Aonori". These are mainly a mixture of algae belonging to the genera *Ulva, Enteromorpha* and *Monostroma* (Hitoegusa).

Sun drying is also practiced in Korea. In the Wando archipelago (South Korea), algae are dried not on sand but on huge nets. In Chile, Gracilaria (*Gracilaria gracilis*) is also sun-dried on wires or racks (see Figures 5.2 and 5.3) (Perez 1997).

This procedure limits contact with the ground and therefore contamination problems. It also facilitates drying by ventilation, particularly in the case of suspended wires.

Figure 5.2. *Drying* Gracilaria *on wires in Chile (photo credit © Barbaroux O. 1997)*

5.2.2. *Ash drying*

Ash drying is a variation on the traditional sun drying method. It is a traditional practice that precedes the process of dehydrating the algae in the open air. In this technique, the algae are mixed in a mixer containing fern or couch grass ash (see Figure 5.4). It is then spread out in the sun in a location chosen for its constant airflow to ensure even drying (see Figure 5.5).

Figure 5.3. *Drying* Gracilaria *on racks in Chile (photo credit © Barbaroux O. 1997)*

Figure 5.4. *Mixing algae in fern ash (Japan) (photo credit © Barbaroux O. 2016)*

By reducing the acidity of the algae tissue and limiting the action of tissue-degrading enzymes (proteases), the ash means the algae can retain a soft, non-brittle texture after dehydration. This ash coating technique, which precedes sun drying, is also applied to another Laminaria species, *Undaria pinnatifida*. The resulting product is marketed under the name "Haiboshi-Wakame", with an annual production of approximately 10,000 tons (Nisizawa et al. 1987).

Figure 5.5. *Sun drying of algae (Japan) (photo credit © Barbaroux O. 2016)*

5.2.3. *Drying by thermal heating*

This rapid dehydration method replaced the method of drying algae in the open air and in the sun in Europe, particularly in France. Algae harvesters were required to collect 3 tons per day, to which was added the time needed to spread out and turn the harvest on the dunes (Perez 1968).

Thermal drying was therefore introduced in response to an industrial constraint caused by a lack of labor for harvesting and processing algae, which is not the case in most Asian countries. However, hot air drying is used in Asia, particularly for the production of Nori sheets (*Porphyra* spp.), which are used in the production of maki rolls (see Figure 5.6).

However, this drying method must meet two requirements. It must be economically viable and only slightly increase the cost price of the algae raw material. It must also have as little impact as possible on the quality of the raw material and the products derived from it, such as preserving the physicochemical properties of phycocolloids (Chapter 6, section 6.7). The drying time and temperature are therefore two important parameters for meeting these economic and quality requirements.

Figure 5.6. *Dried Nori sheets (photo credit © see: https://www.flickr.com/photos/davidandalina/2014893638/)*

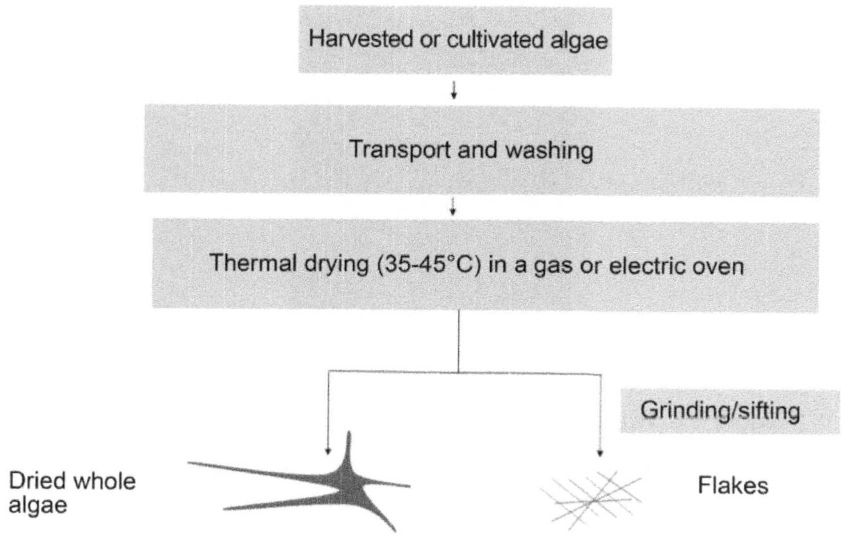

Figure 5.7. *Example of unit processing of algal raw material including a thermal drying step (tobacco dryer, electric oven)*

Most algae intended for human consumption are dried using a very gentle heating process in order to best preserve the product's texture and nutritional quality. After washing, the algae are placed on racks which are then placed in closed chambers similar to those used for tobacco drying.

This technique, known as "tobacco drying", is widely used and reasonably inexpensive. It is based on the circulation of heated air between 35 and 45°C from top to bottom. This hot air is produced by gas combustion. However, there is an alternative that uses electric heating instead of gas.

The heating time varies depending on the species and is often a closely guarded industrial secret. However, it is determined by the final water content of the algae, which must generally be 15% of the product obtained.

Thermal drying is a process frequently used in Europe in the production of dried or flaked algae for human consumption (see Figures 5.7 and 5.8).

Figure 5.8. *Wakame (*Undaria pinnatifida*) dried by thermal heating and marketed in Europe (photo credit © Fleurence J. 2024)*

5.3. Appertization

Appertization is a well-known process that has been mastered by the food industry. It is already used for many algae products and has therefore quickly become established. This process, known to the general public as "preserves", is even subject to a regulatory definition (Article 2 of French Decree No. 55-241 of February 10, 1955):

> Foodstuffs of plant or animal origin that are perishable and whose preservation is ensured by the combined use of the following two techniques are considered to be preserves:
>
> 1. Packaging in a container that is impervious to liquids, gases, and microorganisms at any temperature below 55°C.
>
> 2. Treatment by heat or any other method authorized by decree [...]. The purpose of this treatment must be to completely destroy or inhibit, on the one hand, enzymes and, on the other hand, microorganisms and their toxins, whose presence and proliferation could alter the foodstuff in question or render it unfit for human consumption.

Currently, the appertization process is based on heating the product under pressure. The temperatures used vary between 110 and 130°C, and the minimum application time for these temperatures is 20 minutes. The pressure applied varies (2–4 bars) depending on the product and often needs to be adjusted according to the final specifications. This last piece of information is not easily accessible, as it is part of industrial know-how and often remains confidential.

The species *Laminaria digitata* (Breton Kombu) was initially produced in this form in the late 1980s. Currently, another species, *Himanthalia elongata*, known as "sea spaghetti" or "sea beans", is mainly processed using this method (Box 5.1, Figure 5.9).

Sea spaghetti, or sea beans, are mainly sold in canned form. They are produced under various brand names and are readily available in some supermarkets and specialty organic stores. On store shelves, they may be found alongside canned salicornia, which can cause confusion among some consumers.

Figure 5.9. *Example of a canned or preserved product, the alga* Himanthalia elongata *(photo credit © Fleurence J. 2025)*

Box 5.1. *France is one of the first countries to have developed appertization for edible algae*

5.4. Salting and brining

Salting is more akin to a preservation process than an industrial processing procedure. It has been practiced for a long time to protect foodstuffs from bacterial degradation. Salting is an operation that involves coating the product to be preserved with salt. The amount of salt used is approximately 15% of the weight of the product to be salted.

Brining consists of immersing the product in a liquid with a high salt concentration, sometimes containing herbs and various ingredients. This procedure is therefore a preservation operation, as well as a processing operation, since brining alters the initial product's organoleptic qualities.

These two traditional techniques are also used to preserve and even transform edible algae.

Salting applied to algae is a highly developed technique in Asia. It is used to stabilize the raw material and preserve it from contamination by microorganisms. It is a traditional process used to preserve many edible algae in Japan. This procedure is mentioned in a bulletin from the US Congress Fisheries Office (Smith 1905). In this document, salting is described as a technique for preserving "Mozuku" and "Somen-Nori" algae. The first common name mainly covers brown algae belonging to the genera *Cladosiphon* or *Nemacystus*. The second refers to red algae, but its taxonomic identification remains to be confirmed.

Salting is a process that is also used to stabilize the main edible algae consumed worldwide and in Japan, namely *Laminaria japonica* now renamed *Saccharina japonica*. This alga, known by the common name "Kombu" or "Konbu", is also sold to consumers in salted form. In this form, the product is marketed under the name "Shio-Kombu" (Nisizawa et al. 1987). In Japan, Shio-Kombu production accounted for nearly 11% of the tonnage of Kombu intended for human consumption in the early 1960s. These tonnages then declined over time (see Figure 5.10) (Nisizawa et al. 1987).

Salting was also a process used to transform the alga *Undaria pinnatifida* (Wakame) frequently used in Korea from 1970 to 1990 (Hwang and Park 2020). This process can also be a step in more complex processing methods. This is particularly the case for the process used to obtain salted blanched Wakame, known as "Judoshi-Wakame" (see Figure 5.11). Salted blanched Wakame can also be a starting product for the "cut Wakame" processing method (see Figures 5.12 and 5.13).

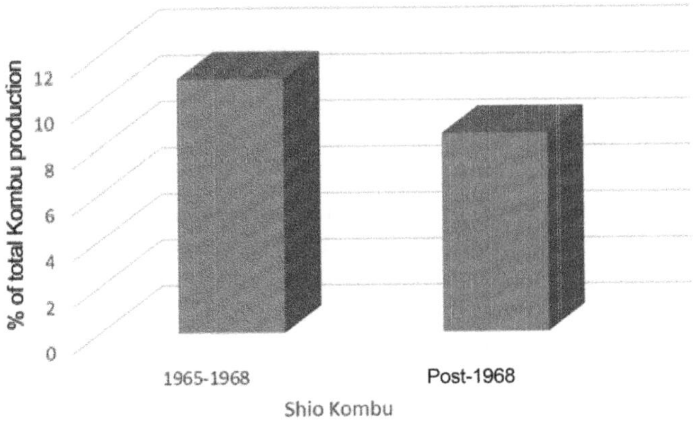

Figure 5.10. *Evolution of salted Kombu (Shio-Kombu) production in Japan in the 1960s (Nisizawa et al. 1987)*

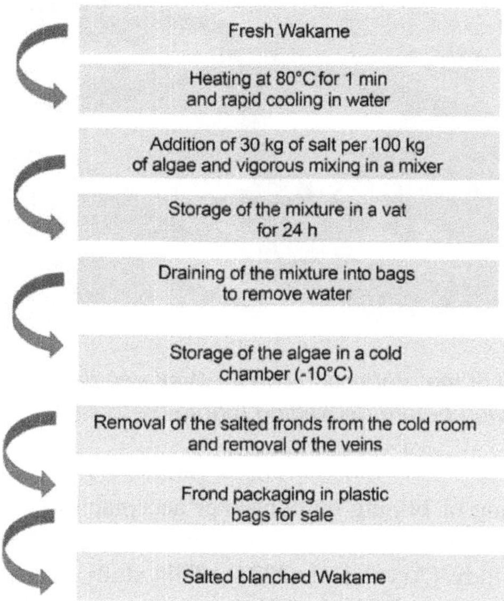

Figure 5.11. *Process for obtaining salted blanched Wakame (Judoshi-Wakame) (Nisizawa et al. 1987)*

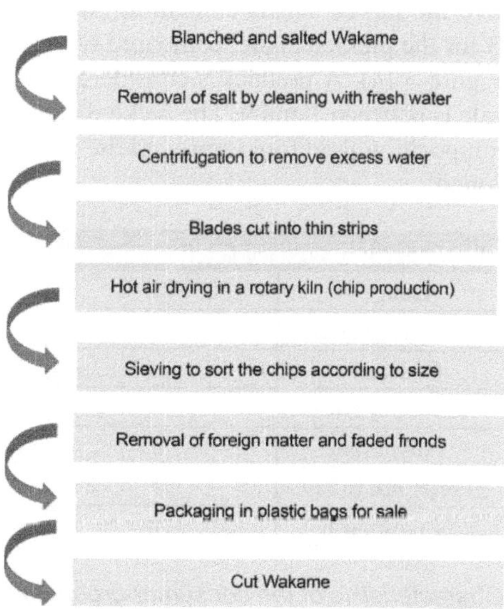

Figure 5.12. *Harvesting process for cut Wakame (Nisizawa et al. 1987)*

Figure 5.13. *Detail of the production line for Wakame flakes (cut Wakame) after cutting and sieving before packaging (photo credit © Barbaroux O. 2016)*

The effect of salting or brining on consumer acceptability of the alga *Saccharina latissima*, formerly known as *Laminaria saccharina* (sea belt) or sugar kelp, was the subject of a recent study (Arya et al. 2024). This study is interesting because the sample group used was not limited to Asian consumers, who are accustomed to consuming algae in various forms (Table 5.3). It included approximately 40 volunteers aged 18–65, broken down by gender as follows: 39% men, 56% women and 4.9% people identifying as non-binary.

The same group gave the mixed salads containing salted algae a score of 6 for the salted algae and 6.3 for the pickled algae, compared with an overall score of 5 for the control salad (see Figure 5.14). A product is considered very acceptable when its value on the hedonic scale is between 7 and 9. The pickled algae used as an ingredient in the proposed food composition therefore improved the taste perception among the consumer group questioned.

Ethnic origin	Number
Asian	7
African American	8
Caucasian	25
Native American	0
Polynesian	0
Prefer not to say	1
No response	1

Table 5.3. *Characteristics of the consumer group questioned for the study of the acceptability of salted or pickled products based on* Saccharina latissima *(Arya et al. 2024)*

In the West, and particularly in France, salting and brining are mainly used as a means of preserving edible algae. Algae that have undergone this treatment are sometimes marketed as "fresh algae", which excludes any further industrial processing.

The fresh alga *Alaria esculenta*, known by the common name Atlantic Wakame, cultivated and sold in the United States and Europe, may undergo a salting process to stabilize it before sale. This alga shows no change in microbiological quality after 90 days of refrigerated storage (5°C) and after prior salting (50 g of NaCl/kg of algae) (Perry et al. 2019). Organoleptic analysis of algae-based salads shows a consumer preference for products with a high salt content (180 g/kg vs. 50 g/kg). This finding is consistent with that previously reported for *Laminaria saccharina* by Ayra et al. (2024).

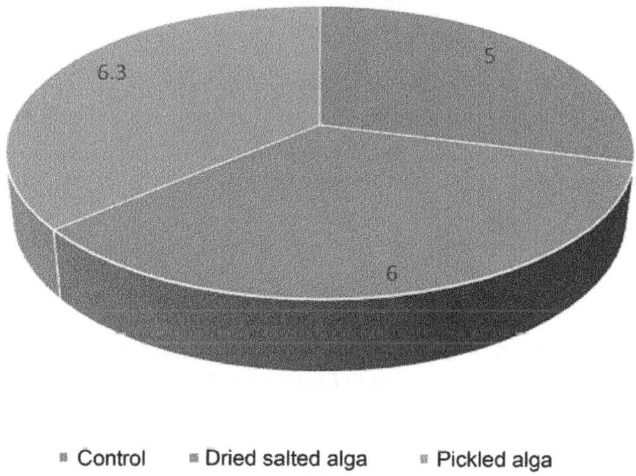

Figure 5.14. *Acceptability of salted (30% weight/weight) or brined (40% weight/volume) algae products in mixed salads according to the hedonic scale compared to a control salad without algae (Arya et al. 2024)*

5.5. Fast freezing

Fast freezing is a classic food preservation process. It is traditionally used for plant-based foods, including algae. This industrial process involves exposing food to very low temperatures (–30 to –50°C) for a very short period of time in order to preserve its nutritional and organoleptic qualities as much as possible. Edible algae are also subject to this cold treatment and preservation process. Fast freezing can be applied to "raw" algae after washing or to algae that have been blanched beforehand.

The effect of fast freezing and freezing on the quality of algae has been the subject of few scientific publications. However, a study conducted on *Undaria pinnatifida* showed that the best fast freezing temperature for preserving the color, aroma and texture of Wakame was –30°C. Furthermore, storage at this temperature is optimal if the algae are stored in bags containing seawater (50% w/v) (Choi et al. 2012).

Although it is an effective preservation method, fast freezing is not optimal for Wakame unless it is combined with storage in seawater. In Europe, fast frozen Wakame is used to make salads with sesame seeds (see Figure 5.15).

Figure 5.15. *Frozen Wakame used for salad production (photo credit © Fleurence J. 2024)*

Another product, "Mozuku", composed of brown algae (*Cladosiphon okamuranus*, *Nemacystus decipiens*) and frequently consumed in Japan, is also frozen before being sold, particularly for export. It is sold as a salad of marinated and frozen algae called "Ajitsuke mozuku" (see Figure 5.16(a)–(c)). It is also known to the Japanese population as "angel hair" due to the filamentous appearance of the algae that makes up the product (see Figure 5.16(c)). Mozuku is mainly harvested in the Okinawa archipelago (Omoto et al. 2024), and some authors suggest that its consumption contributes to the now famous longevity of the island's inhabitants. Local production has declined sharply since 2007 (see Figure 5.17). It is sold in Japan as a fresh, dried or salted product. It can also be sold frozen for export.

Fast freezing is a process that is not only used to preserve algae as food. It can also be used to preserve a stock of individuals ready for cultivation during the winter period. This is particularly the case for the cultivation of the red alga *Porphyra yezoensis* (Nori), now known as *Pyropia yezoensis*.

a)

b)

c)

Figure 5.16. *Frozen Mozuku alga (e.g.* Cladosiphon okamuranus*) or angel hair (photo credit © Fleurence J. 2024)*

In order to prevent the collapse of *Porphyra* production from mid- to late November, stocks of individuals previously preserved by freezing are brought to the cultivation areas. This additional resource is obtained through fast freezing of nets containing algae measuring 1–3 cm. These nets are initially dried in the open air until the thallus water content is between 20 and 40% of the product. When this level of dehydration is reached, the nets are packed in plastic bags, deep-frozen and then stored at –20°C (Oohusa 1993).

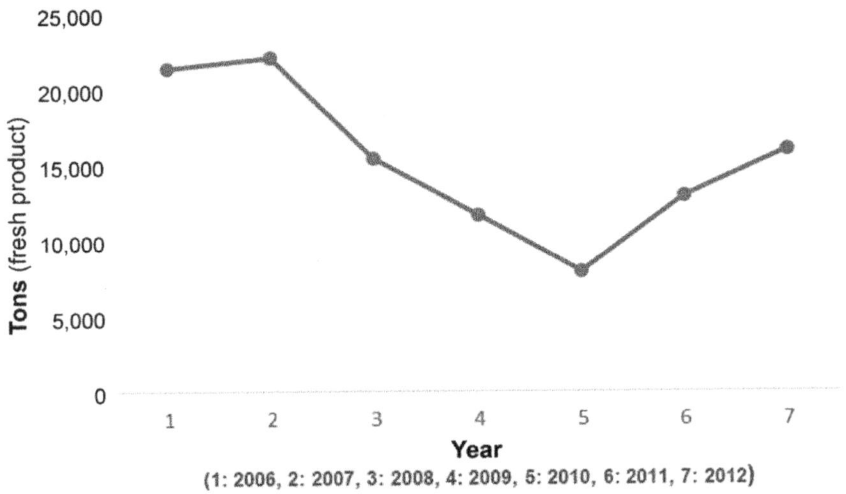

Figure 5.17. *Changes in Mozuku production in the Okinawa archipelago from 2006 to 2012 (Omoto et al. 2024)*

5.6. Freeze-drying

Freeze-drying is a process that involves a fast freezing stage followed by a sublimation stage, resulting in the final dehydration of the product. This process is also known as cryodrying. During fast freezing, the water contained in the product is frozen into ice, which is then evaporated under vacuum through the process of sublimation. This is a very effective method of processing and preserving the product, but it is costly from an economic standpoint.

The application of freeze-drying to algae has many advantages, the main ones being the preservation of nutritional value and flavor, as well as an increase in the product's shelf life. Another advantage, which is of economic interest, is the reduction in the food's weight and volume. This is beneficial in terms of reducing storage and transport costs (see Figure 5.18) (Ashworth et al. 2024).

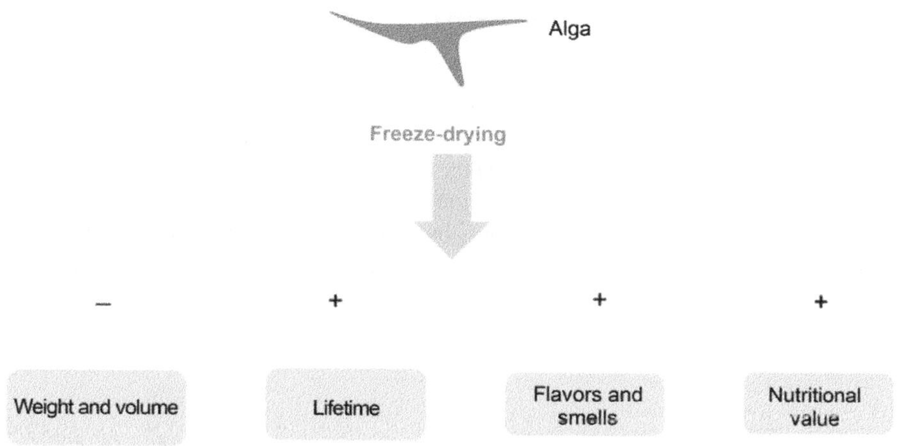

Figure 5.18. *Impact of freeze-drying on the quality and characteristics of the final product*

5.7. Fermentation

Fermentation is a biological process based on the use of microorganisms that break down sugars in an oxygen-poor environment. There are several types of fermentation depending on the biological agents involved and the end products obtained (alcohol, lactic acid). When the product is ethanol, it is called alcoholic fermentation, and when the end product is lactic acid, it is called lactic fermentation. Lactic fermentation, also known as lactofermentation, relies on the use of bacteria or lactic ferments.

Fermentation, a traditional food processing method, is used to produce many everyday products for human consumption (yogurt, cheese, bread, wine). Algae, like many other products, have been subjected to this process, which has often been tested for feasibility on a laboratory scale and, to a lesser extent, on an industrial scale.

Fermentation tests were carried out on algae, particularly on the red alga *Palmaria palmata* (Dulse). These tests did not use bacteria, but rather micromycetes of the genera *Rhizopus*, *Aspergillus* and *Trichoderma* (Marrion et al. 2003). One of the objectives of this work was to demonstrate the feasibility of fermentation on a red edible alga using a common mold, such as *Rhizopus microsporus*, the variety *chinensis*, which is already used for soybean fermentation. The other objective was to evaluate the impact of the action of fungal strains *Trichoderma pseudokoningii* and

Aspergillus oryzae on improving the digestibility of algal proteins and therefore their nutritional value (Chapter 4).

Inoculating the algae with fungal strains significantly improved the digestibility of *Palmaria palmata* proteins (see Figure 5.19). This was due to the degradation of algal polysaccharide fibers (dietary fiber) by fungal enzymes, particularly by xylanase from *Trichoderma reseii*. This degradation of sugars, which is the basis of the fermentation process, was carried out by the three fungal strains tested. It demonstrates the value of this process for improving the nutritional value of *P. palmata* algal proteins.

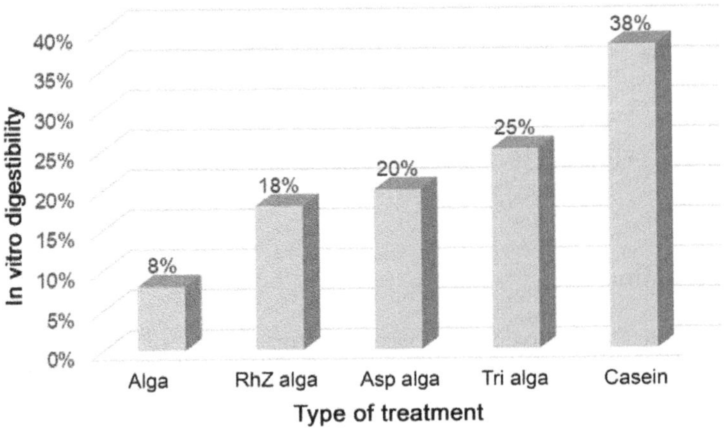

Figure 5.19. *Effect of inoculating the alga* Palmaria palmata *with the fungal strains* Rhizopus microsporus *var.* chinensis *(RhZ),* Aspergillus oryzae *(Asp),* Trichoderma reseii *(Tri) on the digestibility of the alga's protein fraction (% of N released) (Marrion et al. 2003)*

The fermentation of edible algae has also been tested for ethanol production. This was particularly the case for the alga *Undaria pinnatifida*, which was subjected to the fermentative activity of the yeast *Pichia angophorae* under different salt concentration conditions (Cho et al. 2013). The best yield in terms of ethanol production was 9.42 g.L^{-1}. This was observed at a temperature of 30°C, a NaCl concentration of 90 psu, and a fermentation time of 72 hours.

The same yeast strain was used on another brown alga species used for food, *Saccharina japonica* (Dasima or Dashima in Korean) (Jang et al. 2012). In this species, the fermentation process (alcohol production) was coupled with saccharification (enzymatic production of simple sugars from complex sugars). This simultaneous saccharification and fermentation process, based on the use of a bacterial strain (*Bacillus* sp. JS1) and a yeast (*Pichia angophorae*), led to ethanol production of 7.7 g.L^{-1} after 64 hours of fermentation.

In this process, bacterial saccharification appears to be a pretreatment for fermentation, facilitating the fermentation activity of yeast by releasing reducing sugars from algal polysaccharides (see Figure 5.18). It is more effective than other pretreatments based on hot acid hydrolysis or a combination of acid hydrolysis and enzymatic hydrolysis (termamyl) (see Figure 5.20).

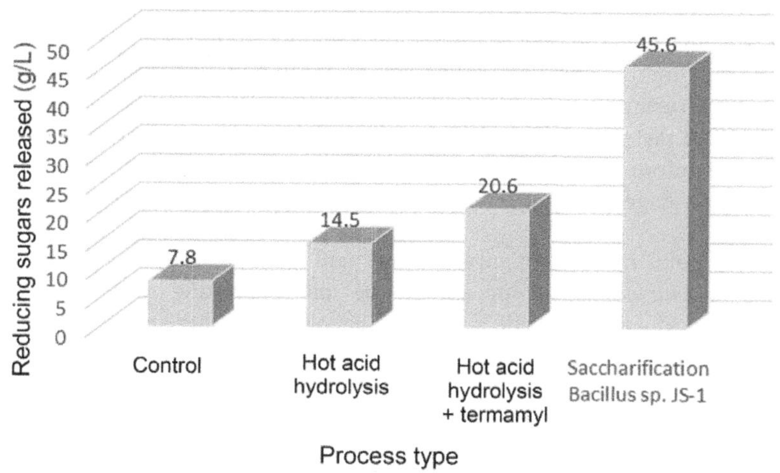

Figure 5.20. *Effect of different pretreatments on the efficiency of the fermentation process of the alga* Saccharina japonica *(Jang et al. 2012)*

Alcoholic fermentation, whether or not coupled with distillation, can integrate algae into the production of certain products (gin, sake, beers) (Kraan 2016). This is currently one of the main industrial food applications of alcoholic fermentation of algae.

The lactic fermentation of algae, or lactofermentation, is another processing method that can be used for edible algae. The presence of indigenous microbiota with lactic and alcoholic fermentation activity has been observed in certain species. This is particularly the case for algae belonging to the genus *Ulva* sp. (Uchida and Murata 2004). In these algae, the fermentation process appears to depend on the presence of a lactic bacterium, *Lactobacillus brevis*, and a yeast, *Debaryomyces hansenii*. However, it is initiated by enzymatic pretreatment of the algae using cellulase. This enzyme probably plays a role in initiating the saccharification process.

This study therefore shows that mixed (alcoholic, lactic) fermentation is possible, thus enriching the prospects for the valorization of green algae in particular. It also contributes to the characterization of a microbiota and is likely to be useful for the fermentation of other algae of food interest.

The use of cellulase as a starting point for saccharification, which improves the fermentation process in algae, as well as the possibility of achieving mixed fermentation, has now been demonstrated in other species belonging to the classes Rhodophyceae or Phaeophyceae (Uchida and Miyoshi 2013) (Table 5.4).

However, the effectiveness of lactic and alcoholic fermentation varies between species, as it is highly dependent on the nature of the polysaccharides present in the algae. Lactic fermentation appears to be very effective in Ulva and Zostera. This is probably due to the predominant presence of cellulose or related polymers in these plants and therefore to the increased effectiveness of cellulase action during the saccharification stage.

From the point of view of mixed lactic acid fermentation, a red alga such as *Gracilaria vermiculophylla* appears to be an interesting candidate for the simultaneous production of lactic acid and ethanol. Most other species preferentially produce lactic acid or ethanol using this process. The nature of the biological material therefore strongly influences the biochemical nature of the products resulting from this fermentation process (Table 5.4).

Algal species	Lactic acid (g/100 mL)	Ethanol (g/100 mL)
Gracilaria incurvata	0.25	0.12
Gracilaria vermiculophylla	0.31	0.23
Hypnea charoides	0.22	0.16
Hizikia fusiformis	0.01	0.24
Laminaria japonica (*Saccharina japonica*)	0.16	0.15
Undaria pinnatifida	0.23	0.38
Ulva sp.	0.76	0.16
*Zostera marina**	1.14	0.26

* Marine vascular plant.

Table 5.4. *Production of lactic acid and ethanol by saccharification (cellulase) and fermentation (*L. casei, D. hanseii, Candida *sp.) from different algae or marine plants (Uchida and Miyoshi 2013)*

Lactic fermentation has also been described as a process not for transforming but for preserving the microbiological quality of algae, particularly in fermented algae-based beverages. This has mainly been described using fermented products obtained from the red alga *Gracilaria fisheri* (Prachyakij et al. 2008). The use of the lactic acid bacterium *Lactobacillus plantarum* DW3 as a starter culture during the algae fermentation process gives the final product antibacterial properties against many pathogenic germs (Table 5.5). The term "partial sterilization" is also used at the end of this process. These antimicrobial properties appear to be directly related to the lactic acid content generated by each type of fermentation process (Table 5.6).

Bacterial strain	Inhibition range (mm)			
	Fermentation process + *L. plantarum*	Fermentation process + partial chemical sterilization + *L. plantarum*	Fermentation process + partial chemical sterilization	Control fermentation process
Bacillus cereus	15.0	16.0	14.0	14.0
Staphylococcus aureus	13.0	12.0	8.1	7.2
Escherichia coli	13.2	11.8	8.1	7.1
Salmonella typhi	14.1	7.0	11.9	7.9
Vibrio parahaemolyticus	17.2	19.2	11.2	10.3

Table 5.5. *Antibacterial effect of* Bacillus plantarum *inoculum introduced into the fermentation process of* G. fisheri *(Prachyakij et al. 2008)*

As with many food products, fermentation appears to be a useful processing method for preserving edible algae. However, fermentation processes applied to algae are limited by the nature of the sugars present in algae, some of which are fermentable and others not, such as alginic acids. This means that not all algae species have the same type of fermentation response and this response is also highly dependent on the microorganism used.

A study comparing the relative efficiency of different *Lactobacillus* strains in producing lactic acid from various plant sources, including algae, showed the high variability of the fermentation process applied to algae (Hwang et al. 2011).

Lactobacillus plantarum appears to be the strain with the highest yield for lactic acid production from the various plant sources tested (Table 5.7). Despite this, algae are less sensitive to the lactic fermentation effect of this strain than other plant resources.

	Fermentation process + *L. plantarum*	Fermentation process + partial chemical sterilization + *L. plantarum*	Fermentation process + partial chemical sterilization	Control fermentation process
Ethanol ($g.L^{-1}$)	1.57	0.78	1.55	1.84
Lactic acid ($g.L^{-1}$)	7.48	6.22	5.86	3.72
Acetic acid ($g.L^{-1}$)	2.24	1.81	1.85	2.41

Table 5.6. *Production of organic acids and alcohol by different fermentation treatments applied to* G. fisheri *(Prachyakij et al. 2008)*

	L. brevis	*L. casei*	*L. plantarum*	*L. salivarus*	*L. collinoides*
Wheat	77.0	71.1	86.6	85.3	20.9
Maize	64.6	66.3	80.2	56.6	22.7
Aspen wood	69.0	67.7	82.2	68.1	23.6
Ulva pertusa	41.4	38.9	49.4	29.4	43.6
Laminaria sp.	42.0	46.2	50.1	48.6	30.7
Gelidium amansii	31.6	54.9	50.4	49.1	4.6

Table 5.7. *Production yield (%) of lactic acid from different* Lactobacillus *strains depending on the plant raw materials (Hwang et al. 2011)*

Fermentation, whether alcoholic, lactic or mixed, remains a process that can be applied to edible algae, but its industrial use requires numerous optimization operations. In particular, it must be developed species by species and take into account numerous parameters such as the appropriate choice of microorganisms best suited to the degradation of a complex and not always well-understood polysaccharide matrix. Despite this difficulty, a few food products (drinks, sea vegetables) derived from algae fermentation are currently available (Chapter 6, section 6.1).

5.8. Enzymatic liquefaction or enzymatic hydrolysis

Enzymatic liquefaction is a process of breaking down organic matter using cellulase enzymes, pectinase, with or without proteases. It is often used on plant-based raw materials and is particularly useful for clarifying fruit juices.

Enzymatic liquefaction is based on the degradation of plant polysaccharides, their structural modification and the change in their functional properties (McClearly 1986). This operation results in a change in the texture of the initial product, which goes from a solid to a liquid state, hence the generic term "liquefaction" to describe this process.

The enzymatic digestion process for algae has mainly been described and developed on a laboratory scale, but it can be extrapolated to an industrial scale with certain adaptations. These mainly concern the use of commercially available enzyme preparations that are authorized for food applications. At the laboratory scale, enzymatic liquefaction of algae relies on the use of commercial enzyme preparations, as well as the use of purified or genetically engineered enzymes.

The first work on the enzymatic liquefaction of algae focused on the red alga *Palmaria palmata* (Dulse) (Lahaye and Vigouroux 1992). This alga was treated with commercial enzyme preparations (Celluclast®, Novozym 188®) and an endo-β-1,4-D-xylanase purified from a commercial preparation called "Pentosanase" produced by the fungal strain *Aspergillus niger*. The "Pentosanase" preparation mainly contains xylanase activity, xylans being the main parietal polysaccharides of the alga *P. palmata*.

When used on dried algae, the action of purified xylanase appears to be less effective than that of the commercial enzyme preparation "Pentosanase", which mainly contains xylanase activity as well as having additional activities whose nature is not specified by the supplier. The combined action of the three commercial preparations does not show any synergistic effect, as the enzyme cocktail is less effective than the "Pentosanase" preparation alone. Liquefaction is also observed in the presence of seawater and in the absence of enzyme (Table 5.8).

Enzyme preparation	Liquefaction rate
Seawater without enzyme	62.6%
"Pentosanase"	87.1%
"Pentosanase" + Celluclast + Novozyme 188	82.9%
Purified xylanase	81.2%

Table 5.8. *Enzymatic liquefaction of the alga* Palmaria palmata *from the initially dried algal material (Lahaye and Vigouroux 1992)*

The effectiveness of the enzymatic process differs on fresh algae, a finding that has since been confirmed on other red algae. Regarding *Palmaria palmata* that has not undergone a dehydration process, the action of "Pentosanase" allows for the liquefaction of 81.4% of the algal matter, rather than 87%. Dehydration therefore appears to be a pretreatment that facilitates the enzymatic liquefaction of the alga.

This type of observation also applies when enzymatic hydrolysis is used to facilitate the production of a soluble protein fraction for food purposes. The use of a purified xylanase from *Trichoderma longibrachiatum* (EC 3.2.1.8 Sigma-Aldrich) significantly improves protein extraction caused by enzymatic digestion of *Palmaria palmata* thallus (Dumay et al. 2013). This algae liquefaction process is more effective on dried material than on fresh algae (see Figure 5.21).

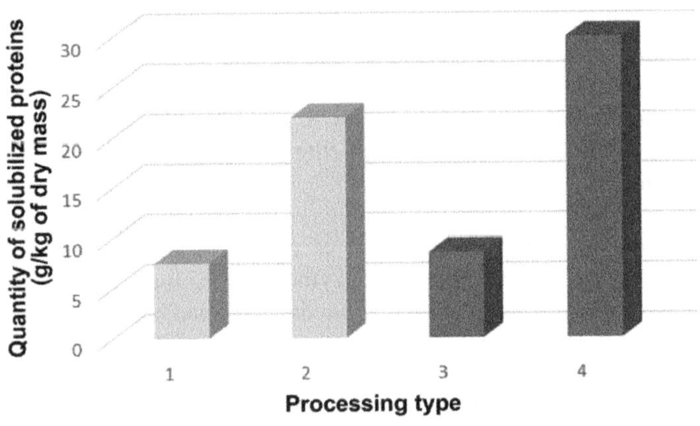

Figure 5.21. *Impact of enzymatic treatment with xylanase on the yield of a soluble protein fraction depending on the state of the algal material (fresh or dried algae) (1: wet algae (control), 2: wet algae + xylanase, 3: dried algae (control), 4: dried algae + xylanase) (Dumay et al. 2013)*

The enzymatic liquefaction process of the alga *Palmaria palmata* was applied to improve the solubilization of algal proteins and their digestibility. It was also used to extract R-phycoerythrin more easily. This purple-pink protein pigment (see Figure 5.22), already used in Asia (Taiwan) as a food coloring, is considered a high value-added molecule. In the biomedical field, its selling price ranges from \$200 to \$220 per mg for the purified molecule.

Figure 5.22. *R-phycoerythrin extracted by enzymatic digestion from the alga* Palmaria palmata *(photo credit © Fleurence J. 2024)*

One of the typical processes used for the enzymatic hydrolysis of *Palmaria palmata* is based on the use of commercially available enzyme preparations. It uses commercial enzymes such as cellulases (Celluclast®, Cellulyve®), xylanases (Feedlyve®, Panlyve®, Shearzyme®) and β-glucanases (Finizym®, Ultraflow®) supplied by Lyven and Novo. This process has also been protected by a patent (Fleurence et al. 2002).

The best results are observed for enzymatic maceration using the enzyme combination Celluclast®-Ultraflow®, which corresponds to the joint use of a cellulase (*Trichoderma reesei*) and a β-glucanase (*Aspergillus niger*) (Table 5.9). The use of xylanases, regardless of their origin, gives results that are inferior to those obtained with β-glucanase preparations. The use of enzymes specific to the degradation of the polysaccharide (xylane), which is mainly present in the cell wall of algae, is therefore less effective than the use of broader-spectrum enzymes such as β-glucanases.

Furthermore, not all xylanases are equally effective. Their effectiveness appears to be highly dependent on the biological origin of the enzymes tested. Indeed, the xylanase obtained from *Aspergillus oryzae* (Shearzyme®) is significantly more effective than that produced by the strains *Trichoderma reesei* (Feedlyve®) and *Aspergillus niger* (Panlyve®) (Table 5.9).

CHARACTERISTICS OF THE SHEARZYME ENZYME COCKTAIL.–

Shearzyme® is produced from a genetically modified strain and its use in food applications requires regulatory approval.

Enzyme preparation	Solubilized protein concentration (mg/L)
Cellulyve® (cellulase)	1,500
Celluclast® (cellulase)	4,000
Panlyve® (xylanase)	4,700
Feedlyve® (xylanase)	5,000
Shearzyme® (xylanase)	8,000
Ultraflow® (β-glucanase)	9,000
Finizym® (β-glucanase)	11,000
Shearzyme® + Ultraflow®	6,000
Shearzyme® + Finizym®	8,000
Shearzyme® + Celluclast®	10,500
Celluclast® + Finizym®	7,000
Celluclast® + Ultraflow®	12,000

Table 5.9. *Effect of different enzyme preparations liquefying the alga* Palmaria palmata *on protein solubilization (Fleurence et al. 2002)*

The use of enzyme combinations gives mixed results, prompting reflection on the use of the enzymatic liquefaction process of *Palmaria palmata* and on the application of this process to other edible algae. It appears that the most effective combination results from the association of activities that do not include xylanase, but less specific activities such as cellulase and glucanase activities (e.g. Celluclast + Ultraflow) (Table 5.9).

However, the combination of "cellulase + xylanase", namely "Shearzyme + Celluclast", appears to be effective in improving the extraction of proteins from *Palmaria palmata*. This combination, used in the pretreatment of the alga, significantly improves the in vitro digestibility of the protein fraction and therefore its potential nutritional value (see Figure 5.23) (Fleurence 2004; Fleurence et al. 2018).

As part of efforts to improve the solubilization of proteins and R-phycoerythrin, the combination of "Shearzyme + Celluclast" was also tested under different operating conditions (different ratios) in order to study the limiting factors of the approach (Joubert and Fleurence 2008).

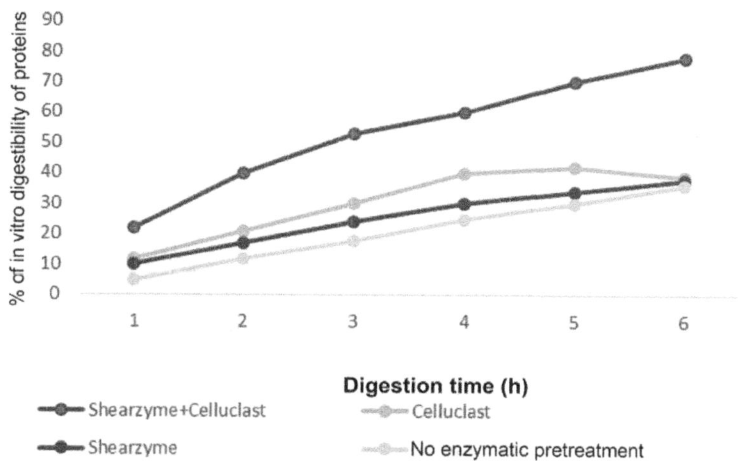

Figure 5.23. *Effect of enzymatic pretreatment (Celluclast, Shearzyme, Shearzyme-Celluclast combination) on the in vitro digestibility of the protein fraction of* Palmaria palmata *(Fleurence 2004; Fleurence et al. 2018)*

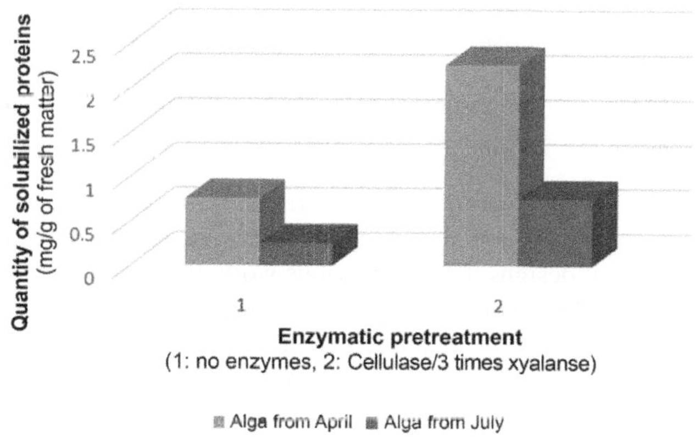

Figure 5.24. *Effect of xylanase concentration on the effectiveness of the "Celluclast + Shearzyme" combination in solubilizing the protein fraction and impact of the seasonal factor (April, July) on the effectiveness of the process (Joubert and Fleurence 2008)*

A concentration of xylanase three times higher than that of cellulase in the combination used increases the amount of protein extracted by enzymatic liquefaction by a factor of 3 (see Figure 5.24). The Shearzyme/Celluclast ratio therefore appears to be a factor limiting the effectiveness of the combination used and, consequently, the process. The different effectiveness levels of the process depending on the season in which the algae are harvested appears to be another determining factor.

This enzyme combination also improves the solubilization of R-phycoerythrin, by a factor of 3–5 compared to the control process (without enzymes) (see Figure 5.25).

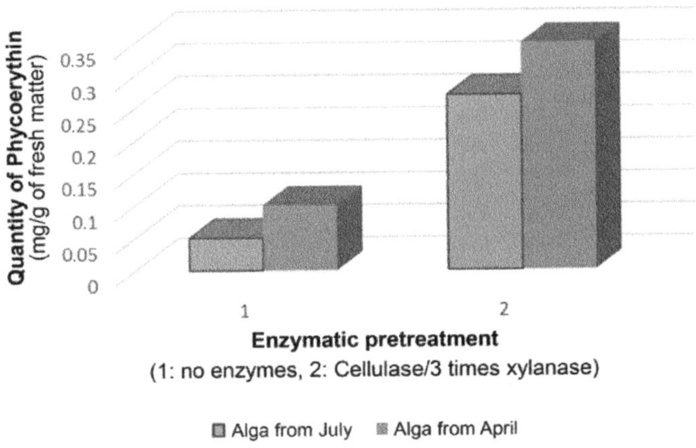

Figure 5.25. *Effect of xylanase concentration on the efficiency of the combination "(Celluclast + Shearzyme)" in the extraction of R-phycoerythrin and impact of the seasonal factor (April, July) on the process's efficiency (Joubert and Fleurence 2008)*

The extraction of R-phycoerythrin (R-PE) from *Palmaria palmata* via enzymatic liquefaction of the alga has been the subject of extensive optimization work based on the use of experimental designs. Unlike previous work, this study is based on the use of purified xylanase rather than commercial enzyme preparations or combinations of activities (Dumay et al. 2013). In particular, it has made it possible to define the optimal conditions for enzymatic liquefaction (temperature, hydrolysis duration, enzyme/alga ratio) and has thus opened up real industrial prospects for obtaining this pigment for use as a food coloring (Table 5.10).

The same work to optimize the liquefaction process using experimental designs was carried out on the red alga *Gracilaria verrucosa*. The objective was the same, namely to develop a gentle extraction process for R-phycoerythrin (Mensi et al. 2012). Unlike the process validated on *P. palmata,* the chosen procedure used a combination of enzymes ("cellulase-xylanase") rather than a purified enzyme. The commercially available cellulase was obtained from *Aspergillus niger* and the xylanase, also commercially available, was produced by *Trichoderma viride.*

	Control (without enzyme)	Liquefaction without optimization	Liquefaction after optimization
Temperature	40°C	40°C	28°C
Hydrolysis time	16.5 min	16.5 min	16.5 min
E/S ratio ($g.kg^{-1}$ DM)	360	360	320
Protein ($g.kg^{-1}$ DM)	8.87	29.80	30.58
R-PE	0.20	3.28	12.36
R-PE purity index	0.03	0.14	0.40

Table 5.10. *Hydrolysis conditions for* Palmaria palmata *via a purified xylanase before and after optimization (Dumay et al. 2013)*

Xylanase is present in small quantities in the enzyme combination used (three times less than cellulase), which is justified by the fact that *Gracilaria verrucosa* is primarily an agarophyte (Table 5.11).

Alga/solvent ratio (mass/volume)	Cellulase rate (mass/volume)	Xylanase content (mass/volume)	Incubation temperature	Incubation time	pH	R-PE extraction yield
20%	15%	5%	48°C	6 h	6.5	6.25 $mg.g^{-1}$ of DM

Table 5.11. *Optimal conditions for the extraction of R-phycoerythrin from* Gracilaria verrucosa *after applying the experimental design approach (Mensi et al. 2012)*

More recently, the extraction of R-PE from the red alga *G. turuturu* by combining enzymatic hydrolysis and sonication has been developed (Le Guillard et al. 2015). This procedure showed that the optimal temperature for R-PE extraction (3.6 mg R-PE.g^{-1} DM at 22°C) was different from that required for optimizing algal liquefaction (91% DM liquefaction at 40°C). This approach reveals that the development of enzymatic liquefaction must be modulated according to the intended application, i.e. improving the extraction of a particular molecule of interest or the processing of a raw material.

The use of enzyme combinations combined with physical treatments, such as sonication or freeze-drying of the algae, remains the preferred approach for optimizing the enzymatic liquefaction process. This combined approach has been successfully applied to red algae such as *Grateloupia turuturu* (Le Guillard et al. 2016) and *Mastocarpus stellatus* (Nguyen et al. 2017).

The enzymatic liquefaction of algae is a processing method whose technical feasibility has now been validated. It requires a preliminary optimization phase because there is no single protocol since the nature of the algal polysaccharides to be degraded varies depending on the species. However, its economic feasibility remains to be demonstrated, and its application to the extraction of high value-added molecules (e.g. natural food coloring) under native conditions remains the best opportunity for the development of this process on an industrial scale.

5.9. Liquid extraction (maceration)

The extraction of soluble compounds by soaking a solid product in a cold liquid is called maceration. This operation can be carried out in water, brine, alcohol or oil. Maceration applied to algae is a simple, widely used process that has been the subject of numerous patents. It is used to prepare extracts enriched in water-soluble or fat-soluble molecules, depending on the nature of the solvent used.

In 1924, an American patent described the maceration of brown algae of the genus *Laminaria* to extract water-soluble compounds, particularly alginates (Thornley et al. 1924). Currently, maceration in an aqueous medium is a common method for extracting alginates from brown algae of the genus *Laminaria*. The industrial process for obtaining alginates comprises five steps: preliminary acidification, liquid extraction in an alkaline medium, followed by a solid/liquid separation step, precipitation of the extracted alginates and a drying operation (Vauchel et al. 2008) (see Figure 5.26).

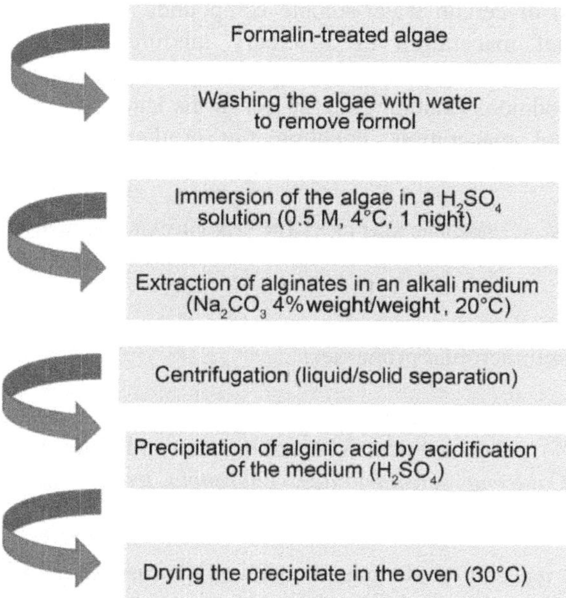

Figure 5.26. *Example of an industrial process used to extract phycocolloids, including a maceration stage of the algae (Vauchel et al. 2008)*

Liquid extraction is carried out by macerating the algae in an alkaline solution of calcium carbonate (4 $g.L^{-1}$) for 16 hours (Perez 1997) or sodium carbonate (see Figure 5.26). In the case of alginates, maceration can be carried out hot or cold. For certain species of Fucales (*Sargassum* sp. *Turbinaria conoides*), hot maceration significantly improves the extraction yield of alginates (Table 5.12). For the alga *Sargassum siliquosum*, hot maceration enables the extraction of 50% of the alginates present in the alga, compared to 39% with the cold method (Chee et al. 2011).

	Cold method	Hot method
Sargassum baccularia	23.9%	26.7%
Sargassum binderi	28.7%	38.7%
Sargassum siliquosum	38.9%	49.9%
Turbinaria conoides	40.5%	41.4%

Table 5.12. *Extraction yield of the alginic fraction according to the extraction method (hot or cold) (Chee et al. 2011)*

The extraction of certain water-soluble compounds, such as fucoidans, may be preceded by brief maceration in a ternary mixture consisting of methanol/chloroform/water (4/2/1) before applying conventional liquid maceration under acidic and/or alkaline conditions (Sinurat et al. 2015). In the latter case, aqueous maceration precedes methanol maceration, enabling the production of extracts that are particularly rich and varied in water-soluble molecules.

ADVANTAGES OF ALCOHOLIC MACERATION FOR OBTAINING BIOLOGICALLY ACTIVE ALGAE EXTRACTS.

Alcoholic maceration can also be used on algae to produce extracts with antioxidant or antimicrobial properties.

Thus, alcohol maceration in methanol solutions of increasing concentrations (50%, 70%, 95%) yields extracts with antioxidant and antiradical activities from algae such as *Caulerpa sertularioides*, *Laurencia tronoi* and *Padina australis* (Dotulong et al. 2013).

The alcoholic maceration of algae to obtain a liquor is also a well-known process. A patent for the alcoholic maceration of the brown alga *Fucus vesiculosus* was filed in the United States (Schirmann 1985). In this process, the algae are first dried in an oven (120°C for 24 hours) and then macerated in 40% or 65% alcohol for three weeks at room temperature. The extract obtained is then filtered. The filtrate can then be mixed with fruit essences, herbs and water before being pasteurized at 85°C for 20 minutes.

Maceration, regardless of the solvent used, remains a simple processing method that is easy to implement and has proven economic feasibility. This process is now widely used for the production of algae extracts, both in the field of food supplements and in intermediate food products used in the composition of beverages in particular.

5.10. Maturation

Maturation is an algae processing method used in Japan for Kombu (Konbu). It involves storing algae, which have been dried in the sun under specific conditions of temperature, darkness and residual humidity. This type of storage triggers a transformation, based on the activation of endogenous enzymes, which modifies the product's physical, chemical and organoleptic characteristics (texture, odor, color, flavor).

The resulting product is called "Kura-gakoi" (see Figure 5.27). It is aged in cellars for between 1 and 10 years. During this maturation process, the product acquires a milder flavor than the strong taste of the original alga, while developing aromatic marine notes. This level of quality is generally achieved after two years of aging (Mouritsen et al. 2012). Kura-gakoi is used in the preparation of the traditional Japanese soup "Dashi", which is a broth made from dried algae and dried bonito.

Figure 5.27. *Dried Kombu matured in an aging cellar, or Kura-gakoi, used to prepare Dashi soup (photo credit © Fleurence J. 2024)*

UMAMI, A FLAVOR GENERATED BY ALGAE.

Kombu Dashi is believed to be the source of the umami flavor so appreciated by Asian gourmets and chefs. This particular flavor, which refers to a fifth taste (salty, sweet, bitter, sour, umami), is thought to be due to the action of glutamate on certain sensory receptors on the tongue.

Recent work on European coastal species, *Saccharina latissima* (Royal Kombu) and *Palmaria palmata* (Dulse), has been carried out to develop "Kombu Dashi" and "Dulse Dashi" in order to enhance the "umami" flavor in new Nordic cuisine.

In northern Europe, particularly in Norway, maturation trials of the algae *Saccharina latissima* (*Laminaria saccharina*) (Kombu or Royal Kombu) and *Palmaria palmata* (Dulse) have been carried out. The algae were first dried, stored in the dark and under maturation conditions that were not specified in detail.

Brown algae of the genus *Laminaria* are generally matured under fairly basic conditions that do not require any special precautions, as the presence of alginates and fucoidans in their thick walls protects them from bacterial or fungal attacks. The presence of a large number of varied pigments, most of which are not water-soluble, also limits the risk of discoloration of the product during maturation.

Red algae, particularly *Palmaria palmata*, are more sensitive to storage conditions during maturation. Their purple pigment, R-phycoerythrin, is highly water-soluble and sensitive to light exposure. For these algae, the maturation process takes place at low temperatures, generally in closed containers protected from light (personal communication, Pierrick Stevant 2017).

Dulse from algaculture appears to be a better source of glutamate than Royal Kombu obtained by the same production method (10 times more glutamate under similar extraction or infusion conditions: 60°C for 45 min). This finding is more nuanced when it comes to harvested Dulse, for which the glutamic acid content is similar to that found in Royal Kombu. As glutamic acid is involved in the development of umami flavor, *Palmaria* palmata (Dulse), regardless of its origin, appears to be an interesting resource for the development of a "Dulse Dashi" broth (Table 5.13).

Broth or Dashi	Glutamic acid (mg/100 g)	Aspartic acid (mg/100 g)	Alanine (mg/100 g)
Rishiri-Kombu Dashi*	22	18	1
Rausu-Kombu Dashi (traditional)	100	60	7
Rausu-Kombu Dashi (vacuum-packed)	145	85	20
Dulse Dashi	40	27	25
Ichiban Dashi*	25	18	4
Chicken broth (China)	14	4	8
Chicken broth (Western)	18	6	11

* Ichiban Dashi and Rishiri-Kombu Dashi preparations are obtained with half the amount of algae per liter of water compared to Dulse Dashi preparations.

Table 5.13. *Glutamic and aspartic acid and alanine contents in two algae-based Dashi soups (Mouritsen et al. 2012)*

The algae maturation process is traditional in Japan but still underdeveloped in the West. It is based on empirical practice that begins with sun drying the algae and is therefore applied to a raw material whose residual moisture content varies depending on the species and region. Furthermore, storage and maturation conditions are rarely disclosed, as they are part of industrial know-how and are very often protected by confidentiality. Despite this, recent work on the maturation of *Palmaria palmata* provides information on the impact of storage conditions on the matured product's organoleptic qualities (Stévant et al. 2020; Fleurence 2022).

Dried algae are stored in the dark at 12°C for up to 126 days. The moisture content of algae subjected to the maturation process is 20% for semi-dry samples and 6% for dry samples. During the maturation process, the content of amino acids involved in the umami flavor varies little during storage for semi-dried samples (20% moisture content) (see Figure 5.28).

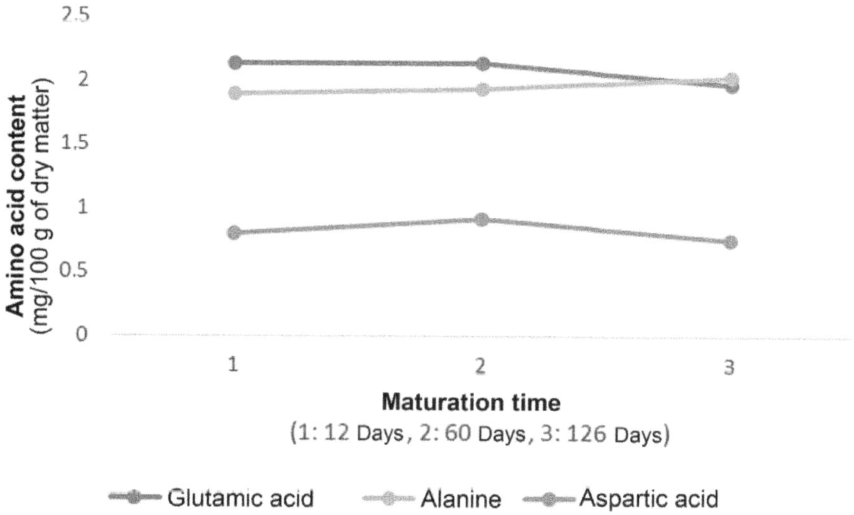

Figure 5.28. *Changes in the levels of amino acids involved in umami flavor during the maturation process of semi-dried Dulse samples (20% moisture content) stored in the dark at 12°C (Stévant et al. 2020)*

However, during the maturation process of dried samples (6% moisture content), a significant decrease in glutamic and aspartic acid contents was observed (see Figure 5.29) Maturation of dried samples does, however, have advantages. In particular, it allows for better preservation of marine flavors and aromas (see Figure 5.30).

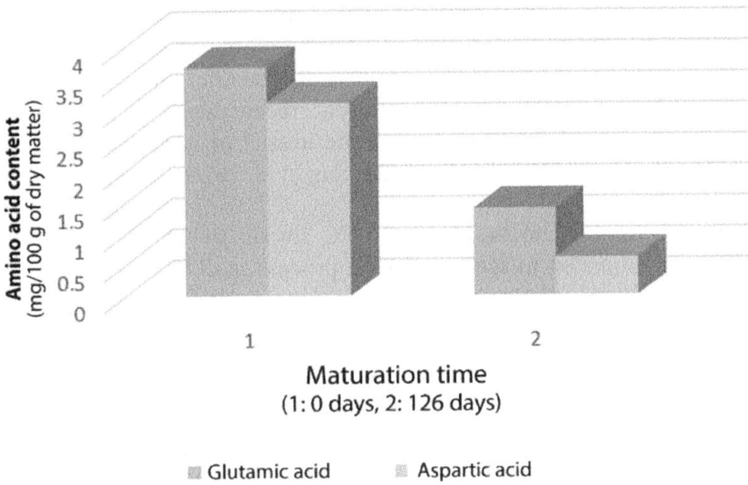

Figure 5.29. *Changes in the levels of amino acids involved in umami flavor during the maturation process of samples of dried Dulse (6% moisture content) placed in the dark at 12°C (Stévant et al. 2020)*

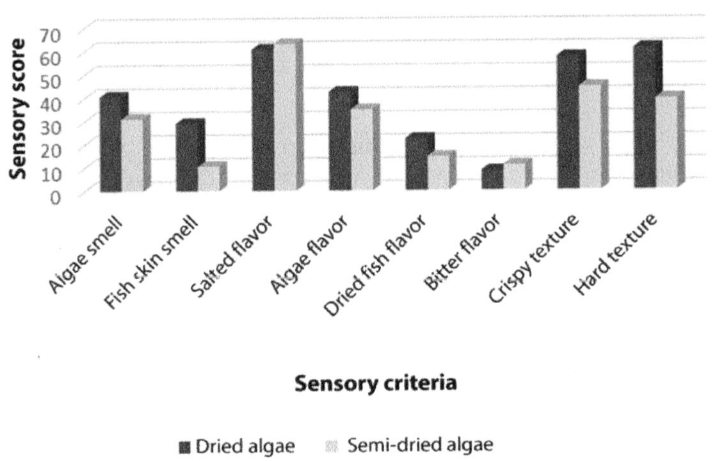

Figure 5.30. *Organoleptic profile after 126 days of maturation of dry algae (6% moisture content) and semi-dried algae (20% moisture content) (Stévant et al. 2020)*

The maturation process carried out under the conditions mentioned above results in a microbiological quality of both products that is satisfactory for food application.

The aerobic flora load is 10.4 x 10^3 germs/g of product for algae dried and matured for 126 days (see Figure 5.31). It is only 5.9 x 10^4 germs/g of product for semi-dried algae after 126 days of maturation. By way of comparison, the maximum mesophilic aerobic flora count permitted in France for edible algae is 100,000 germs/g (Fleurence 2018).

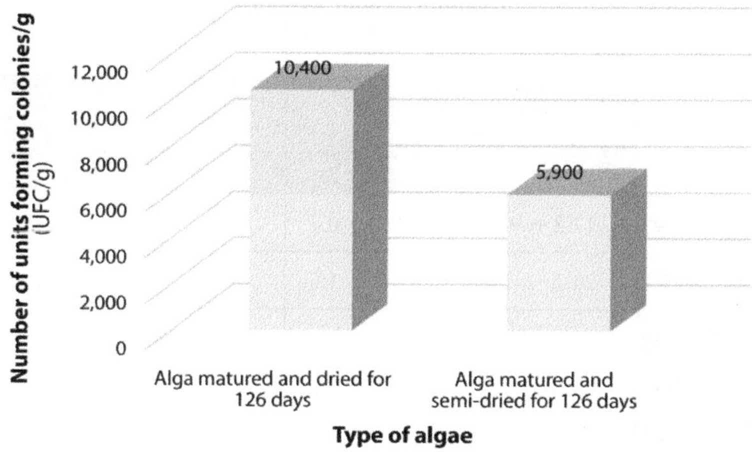

Figure 5.31. *Microbiological load (total aerobic flora) of dried and semi-dried algae after 126 days of maturation (Stévant et al. 2020)*

The absence of fecal coliforms is also noteworthy in both types of products.

5.11. Preservation in seawater

The preservation of algae in seawater is a process that has also been studied, with a particular focus on the storage conditions likely to induce changes in the nutritional value and organoleptic qualities of the raw material subjected to these conditions.

It was recently tested on two brown algae used for food, *Alaria esculenta* and *Saccharina latissima* (Stevant et al. 2017). Storage in seawater at 18°C for a period exceeding 22 hours appears to have an effect on the chemical composition and color of the algae (Table 5.14). Thus, storage under these conditions leads to a loss of dry matter (approximately 3%) for both species considered. Conversely, this storage method leads to an increase in the mineral content (+25%) coupled with a significant increase in the sodium content. A significant decrease in the carbohydrate, polyphenol and fucoxanthin levels of sodium was also observed, while the protein content appeared to be largely unaffected (Table 5.14). A marked variation in thallus color was also reported for both species during storage.

This simple and inexpensive process therefore appears to offer new prospects for preservation. Its impact on the chemical composition and color of the product is real for the algae tested. This process can therefore be considered a transformation procedure whose technical feasibility and economic interest deserve to be investigated further in the context of industrial validation.

	Storage time *Alaria esculenta*		Storage time *Saccharina latissima*	
	0	22	0	22
Dry matter	17.2 +/–0.8	13.9 +/–0.7	16.1 +/–1.3	13.5 +/–0.8
Minerals	24.2 +/–1.4	27.0 +/–1.6	26.2 +/–2.6	30.0 +/–2.1
Sodium	3.9 +/–02	5.2 +/–0.2	3.6 +/–0.2	4.3 +/–0.2
Carbohydrates	40.7 +/–1.5	37.7 +/–1.5	46.1 +/–2.5	40.0 +/–1.0
Polyphenols	3.43 +/–0.08	2.55 +/–0.09	0.69 +/–0.04	0.49 +/–0.04
Fucoxanthin	871 +/–53	829 +/–45	431 +/–19	360 +/–27
Protein	10.5 +/–0.2	9.9 +/–0.1	10.6 +/–0.3	11.6 +/–0.2

Table 5.14. *Effect of storage time in seawater on the chemical composition of the algae* Alaria esculenta *and* Saccharina latissima *(Stévant et al. 2017)*

Many of the processing methods used for algae are conventionally used in the food industry (drying, appertization, deep freezing, fermentation, etc.). As such, algae appear to be an easy raw material to process using unit operations that are traditionally mastered by the industry.

5.12. References

Arya, R., Camire, M.E., Skonberg, D.I., Perry, J.J. (2024). Effect of dry salting and brining on the consumer acceptance of *Saccharina latissima* (sugar kelp). *Phycology*, 4(2), 330–339.

Ashworth, H., Donohoe, P., Kibblewhite, B. (2024). Seaweed and the applicability of freeze drying techniques. *Frontiers in Marine Science*, 11, 1386418.

Chee, S.Y., Wong, P.K., Wong, C.L. (2011). Extraction and characterization of alginate from brown seaweeds (Fucales, Phaephyceae) collected from Port Dickson, Peninsular Malaysia. *Journal of Applied Phycology*, 23, 191–196.

Cho, Y., Kim, H., Kim, S.K. (2013). Bioethanol production from brown seaweed, *Undaria pinnatifida*, using NaCl acclimated yeast. *Bioprocess Biosystem Engineering*, 36, 713–719.

Choi, J.S., Lee, B.B., An, S.J., Sohn, J.H., Cho, K.K., Choi, I.S. (2012). Simple freezing and thawing protocol for long-term storage of harvested fresh *Undaria pinnatifida. Fisheries Science*, 78, 1117–1123.

Dotulong, V., Widjanarko, S.B., Yunianta, L.P. (2013). Antioxidant activity of three marine algae methanol extract collected from north Sulawesi waters, Indonesia. *International Journal of Science and Engineering Investigations*, 23, 26–30.

Dumay, J., Clément, N., Morançais, M., Fleurence, J. (2013). Optimization of hydrolysis of *Palmaria palmata* to enhance R-phycoerythrin extraction. *Bioresource Technology*, 131, 21–27.

Fleurence, J. (2004). Seaweed proteins. In *Proteins in Food Processing*, Yada, R.Y. (ed.). Woodhead Publishing, Sawston.

Fleurence, J. (2018). *Les algues alimentaires : bilan et perspectives*. Lavoisier Tec & Doc, Paris.

Fleurence, J. (2022). Biotechnological processes applied to edible seaweeds: What perspectives?. *Trends in Food Science & Technology*, 129, 617–620.

Fleurence, J., Antoine, E., Luçon, M. (2002). Procédé d'extraction et d'amélioration de la digestibilité des protéines de *Palmaria palmata*. European patent, EP 1301088 B1.

Fleurence, J., Morançais, M., Dumay, J. (2018). Seaweed proteins. In *Proteins in Food Processing*, Yada, R.Y. (ed.). Woodhead Publishing, Sawston.

Hwang, E.K. and Park, C.S. (2020). Seaweed cultivation and utilization of Korea. *Algae*, 35(2), 107–121.

Hwang, H.J., Lee, S.Y., Kim, S.M., Lee, S.B. (2011). Fermentation of seaweed sugars by *lactobacillus* species in the potential of seaweed as a biomass feedstock. *Biotechnology and Bioprocess Engineering*, 16, 1231–1239.

Jang, J.S., Cho, Y., Jeong, G.T., Kim, S.K. (2012). Optimization of saccharification and ethanol production by simultaneous saccharification and fermentation (SSF) from seaweed, *Saccharina japonica*. *Bioprocess Biosystem Engineering*, 35, 11–18.

Joubert, Y. and Fleurence, J. (2008). Simultaneous extraction of proteins and DNA by enzymatic treatment of the cell wall of *Palmaria palmata* (Rhodophyta). *Journal of Applied Phycology*, 20, 55–61.

Kraan, S. (2016). Seaweed and alcohol: Biofuel or booze. In *Seaweed in Health and Disease Prevention*, Fleurence, J. and Levine, I. (eds.). Academic Press, Elsevier, Amsterdam.

Lahaye, M. and Vigouroux, J. (1992). Liquefaction of dulse (*Palmaria palmata* (L.) Kuntze) by a commercial enzyme preparation and a purified endo-β-1,4 D xylanase. *Journal of Applied Phycology*, 4, 329–337.

Le Guillard, C., Dumay, J., Donnay-Moreno, C., Bruzac, S., Ragon, J.Y., Fleurence, J., Bergé, J.P. (2015). Ultrasound-assisted extraction of R-phycoerythrin from *Grateloupia turuturu* with and without enzyme addition. *Algal Research*, 12, 522–528.

Le Guillard, C., Bergé, J.P., Donnay-Moreno, C., Bruzac, S., Ragon, J.Y., Baron, R., Fleurence, J. (2016). Soft liquefaction of the red seaweed *Grateloupia turuturu* Yamada by ultrasound-assisted enzymatic hydrolysis process. *Journal of Applied Phycology*, 28, 2575–2585.

Marrion, O., Schwertz, A., Fleurence, J., Guéant, J.L., Villaume, C. (2003). Improvement of the digestibility of the proteins of the red alga *Palmaria palmata* by physical processes and fermentation. *Nahrung*, 5, 339–344.

McClearly, B.V. (1986). Enzymatic modification of plant polysaccharides. *International Journal Biological Macromolecules*, 8, 349–354.

Mensi, F., Ksouri, J., Seale, E., Romdhane, M.S., Fleurence, J. (2012). A statistical approach for optimization of R-phycoerythrin extraction from the red algae *Gracilaria verrucosa* by enzymatic hydrolysis using central composite design and desirability function. *Journal of Applied Phycology*, 24, 915–926.

Mouritsen, O.G., Williams, L., Bjerregaard, R., Duelund, L. (2012). Seaweeds for umami flavour in the new nordic cuisine. *Flavour*, 1, 4.

Nguyen, H.P.T., Morançais, M., Fleurence, J., Dumay, J. (2017). *Mastocarpus stellatus* as a source of R-phycoerythrin: Optimization of enzyme assisted extraction using response surface methodology. *Journal of Applied Phycology*, 29, 1563–1570.

Nisizawa, K., Noda, H., Kikuchi, R., Watanabe, T. (1987). The main seaweed foods in Japan. *Hydrobiologia*, 151(1), 5–29.

Omoto, R., Uehara, M., Seki, D., Kinjo, M. (2024). Supply chain-based coral conservation: The case of Mozuku seaweed farming in Onna Village, Okinawa. *Sustainability*, 16, 2713.

Oohusa, T. (1993). Recent trends in nori products and markets in Asia. *Journal of Applied Phycology*, 5, 155–159.

Perez, R. (1968). La conservation et le stockage des Laminaires. *Sciences et Pêches. Bulletin de l'Institut des Pêches maritimes*, 171, 1–8.

Perez, R. (1997). *Ces algues qui nous entourent.* Ifremer, Paris.

Perry, J.J., Brodt, A., Skonberg, D.I. (2019). Influence of dry salting on quality attributes of farmed kelp (*Alaria esculenta*) during long-term refrigerated storage. *LWT*, 114, 108362.

Prachyakij, P., Charernjiratrakul, W., Kantachote, D. (2008). Improvement in the quality of a fermented seaweed beverage using antiyeast starter of *Lactobacillus plantarum* DW 3 and partial sterilization. *World Journal of Microbiology and Biotechnology*, 24, 1713–1720.

Schirmann, F. (1985). Process for the obtention of an algae beverage. US Patent number 4524084.

Sinurat, E., Peranginangin, R., Saepudin, E. (2015). Purification and characterization of fucoidan from the brown seaweed *Sargassum binderi* sonder. *Squalen Bulletin of Marine and Fisheries Postharvest and Biotechnology*, 10, 79–87.

Smith, H.M. (1905). The seaweed industries of Japan. *Bulletin of the Bureau of Fisheries*, 24, 135–165.

Stévant, P., Marfaing, H., Rustad, T., Sandbakken, I., Fleurence, J., Chapman, A. (2017). Nutritional value on the kelps *Alaria esculenta* and *Saccharina latissima* and effects of short-term storage on biomass quality. *Journal of Applied Phycology*, 29, 2417–2426.

Stévant, P., Ólafsdóttir, A., Déléris, P., Dumay, J., Fleurence, J., Ingadóttir, B., Jónsdóttir, R., Ragueneau, E., Rebours, C., Rustad, T. (2020). Semi-dry storage as a maturation process for improving the sensory characteristics of the edible red seaweed dulse (*Palmaria palmata*). *Algal Research*, 51, 102048.

Thornley, F.C., Ford, F., Reynard, O. (1924). Process for the utilization of seaweed. US Patent number 1509035.

Uchida, M. and Miyoshi, T. (2013). Algal fermentation – The seed for a new fermentation industry of food and related products. *JARQ*, 47, 53–63.

Uchida, M. and Murata, M. (2004). Isolation of lactic acid bacterium and yeast consortium from a fermented material of *Ulva* Spp. (Chlorophyta). *Journal of Applied Microbiology*, 97, 1297–1310.

Vauchel, P., Kaas, R., Arhaliass, A., Baron, R., Legrand, J. (2008). A new process for extracting alginates from *Laminaria digitata*: Reactive extrusion. *Food Bioprocess Technology*, 1, 297–300.

6

Applications in Human Nutrition

6.1. Background

Algae are a widely used food resource. They can be consumed as a sea vegetable or as an ingredient. In this case, they are directly incorporated into the food process. Consumers are therefore aware that algae are part of their diet.

However, algae can also be indirectly associated with food. Algae are a source of phycocolloids (alginates, agar, carrageenans) with thickening and even gelling properties that are useful in the food industry. These macromolecules are frequently used as technological additives in the production of processed food products. In this type of application, the dietary contribution of algae comes from substances extracted from them and can be described as an indirect contribution.

6.2. Sea vegetables

Algae have been part of the human diet for centuries, even millennia. Remains of meals containing brown algae (*Eisenia bicyclis*, *Sargassum* sp., *Undaria pinnatifida*) have been found in Japan at archaeological sites dating back to 10,000 BCE (Arasaki and Arasaki 1985). In Korea, fragments of algae (*Hizikia* sp. and *Undaria* sp.) have also been identified in fossilized meal remains, also dating from the Neolithic period (Fleurence 2016).

While brown algae are part of the traditional diet of Asian populations, other algae are also consumed by certain human populations as part of their ancestral diet.

For a color version of all the figures in this chapter, see www.iste.co.uk/fleurence/marinealgae.zip

This is particularly true of the inhabitants of French Polynesia (Marquesas Islands), who have long consumed green algae such as *Enteromorpha flexuosa* and *Caulerpa racemosa*. The latter species is still used today by local populations under the name "Imu topua" (Conte and Payri 2002).

In the West, prehistoric and ancient populations of the Mediterranean basin showed little interest in marine plants. However, Xenophon (430–355 BCE), the famous Greek philosopher, refers in his work *Anabasis*, to a decoction of marine plants drunk by warriors during heatwaves (Perez 1997). The situation was different for populations living on the North Atlantic coast. Algae became part of the diet of these populations as early as the Viking period (793–1066 CE). The red alga *Palmaria palmata* (*Rhodymenia palmata*) was consumed in particular by the populations of present-day Norway (Delaney et al. 2016). Until the beginning of the 20th century, algae were consumed in northern European countries (the United Kingdom, Ireland and Scandinavia). Camille Sauvageau quotes Edmond Perrier in his 1920 book:

> For a long time, young stems of Laminaria have been eaten on the coasts of England and Norway under the name of tangle; *Alaria esculenta*, a species of Fucus (basionym *Fucus esculentus*) [see figure 6.1], is the ordinary food of the coastal populations of Ireland, Scotland, Denmark and the Faroe Islands.

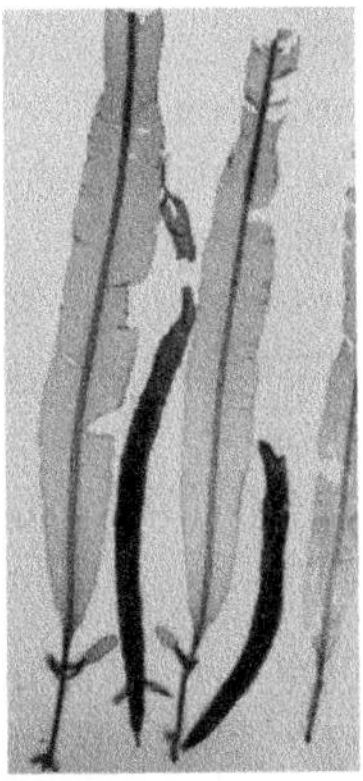

Figure 6.1. *Young specimens of* Alaria esculenta *collected in Finistère (source: personal algae collection J. Fleurence) (photo credit © Fleurence J. 2024)*

During this period, various species were consumed by populations in northern Europe and marketed under various names (Table 6.1). *Palmaria palmata*, a red alga formerly known as *Rhodymenia palmata*, is still known today under three different common names (Figure 6.2 and Table 6.1).

Species	Common name	Country of consumption
Laminaria digitata, L. flexicaulis, L. cloustonii (brown algae)	Tangle	Scotland Ireland
Alaria esculenta A. pylaii (Brown algae)	*Marinkjarni*	Ireland France (Ouessant Island)
Ulva lactuca (green alga)	"Green laver" "Oyster green"	United Kingdom
Palmaria palmata (red alga)	Dulse Dillisk Dilsk	Norway Iceland France (Brittany) Ireland

Table 6.1. *Examples of algae used as sea vegetables in Northern Europe until the beginning of the 20th century (Sauvageau 1920)*

Figure 6.2. Palmaria palmata, *or Dulse (photo credit © Fleurence J. 2020)*

Despite these culinary uses, algae were often associated by coastal populations with substitute foods in times of food shortages or even famine. The advent of potato cultivation in the 17th century ensured food security for populations and definitively relegated algae to the status of "Jerusalem artichoke". This unflattering image of algae as a "food of famine" has strongly influenced the collective unconscious of Western consumers.

The perception of algae as a "sea vegetable" is, however, highly valued in Asia, where they have been part of the food culture for thousands of years. In Japan and Korea, algae are traditionally consumed as sea vegetables and there are many different products available. Brown algae valued as a sea vegetable are used in a wide variety of products. Four groups of algae are mainly used for this purpose: *Laminaria*, *Undaria*, *Hizikia* and *Sargassum*.

Algae of the genus *Laminaria* are marketed under the name Kombu or Konbu. The word "Kombu" means happiness in Japanese. This name reflects the extremely positive image that the Japanese have of algae, which contrasts with the mistrust that Western consumers still have towards these marine plants. The commercial name "Kombu" covers a large number of species, some of which are not of the genus *Laminaria* (Table 6.2).

Species	Kombu	Breton Kombu	Royal Kombu
Laminaria japonica	+		
Laminaria japonica var. *ochotensis*	+		
Laminaria religiosa	+		
Laminaria coriacea	+		
Laminaria angustata	+		
Laminaria angustata var. *longissima*	+		
Arthrothamnus bifidus	+		
Kjellmaniella gyrata	+		
Kjellmaniella crassifolia	+		
Laminaria digitata		+	
Laminaria saccharina			+

Table 6.2. *Brown algae species used as "sea vegetables" under the trade name Kombu (Nisizawa et al. 1987; Fleurence 2016)*

Most products consumed as sea vegetables under the name Kombu are obtained after drying the algae in the sun (Chapter 5, section 5.2.1). Nearly 14 different varieties of products are available to consumers under this name. By way of comparison, there are only two Kombu-type names in France (Breton Kombu and

Royal Kombu), regardless of how the Laminaria resource is processed. In the rest of Europe and the West, algae of the genus *Laminaria* are often marketed under the single, general name "Kelp".

Wakame (*Undaria pinnatifida*) is also consumed as a sea vegetable in both Asia and Europe. In Japan, this product is available under several trade names (Table 6.3). As in the case of Kombu, these names depend on the nature of the processing applied during transformation. However, traditional products are distinguished from new products, as indicated in Table 6.3.

Wakame	Trade name
Wakame, simply dried (1)	Suboshi-Wakame
Wakame, ash-coated and then dried (1)	Haiboshi-Wakame
Wakame dried on a rack (1)	Ita-Wakame
Wakame, blanched and salted (2)	Judoshi-Wakame
Wakame, chopped (2)	Cut Wakame
Sporophyll (whole seaweed) (2)	Mekabu-Wakame
Stipe and central vein (2)	Kuki-Wakame

(1) Traditional product. (2) New product.

Table 6.3. *Different types of Wakame marketed as sea vegetables in Japan (Nisizawa et al. 1987)*

Cut Wakame is a non-traditional product marketed in flake form and used after hydration as a sea vegetable or condiment (Figure 6.3).

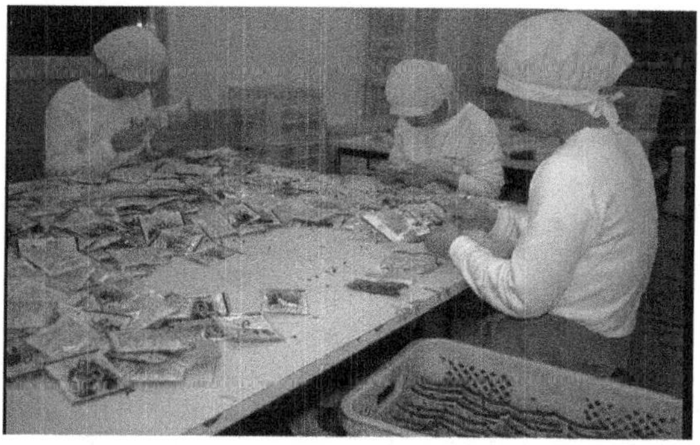

Figure 6.3. *Packaging line for cut Wakame (photo credit © Barbaroux O. 2016)*

Figure 6.4. *Dried Wakame sold in strips for use as a sea vegetable at a Japanese market (photo credit © Barbaroux O. 2016)*

Among the main brown algae consumed in Japan, after Kombu and Wakame (Figure 6.4), Hiziki is the most common. Hiziki is the generic name given to the product derived from the brown alga *Hizikia fusiformis*. This alga, belonging to the family Sargassum, has been consumed for centuries by the Japanese population.

As we have seen previously, there are several types of products on the market that differ in the way they are processed (Chapter 5, section 5.2.1). These products, which are sold as sea vegetables, are listed under the names Suboshi-Hiziki and Hoshi-Hiziki. Suboshi-Hiziki can be used as a sea vegetable, as well as an intermediate food product for the manufacture of more elaborate products. Hoshi-Hiziki is mainly consumed as an accompaniment to boiled soybeans or fried beans.

Apart from Wakame and Kombu, other brown algae used as sea vegetables in human food can be found in Europe.

These are mainly *Fucus* sp. (Figure 6.5), *Ascophyllum nodosum*, but above all *Himanthalia elongata* (Figure 6.6). This species is sold in jars under the name "sea beans" or "sea spaghetti" (Chapter 5, section 5.3).

The species *Alaria esculenta,* known as Irish Wakame, Atlantic Wakame or succulent alaria, is consumed in Europe as a sea vegetable. This alga has traditionally been used as food by coastal populations in northern Europe. In the United Kingdom, it is known by various regional names (Table 6.4). In Ireland and France, it is now the subject of sustained algaculture.

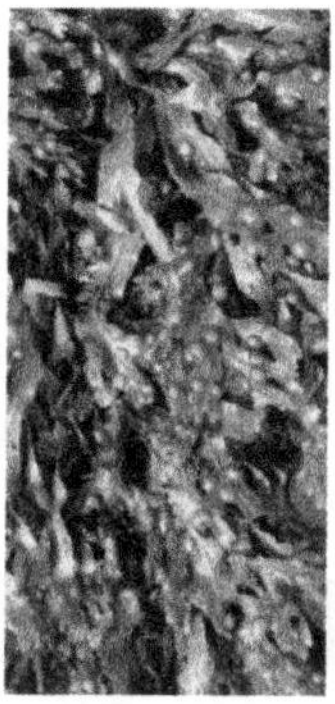

Figure 6.5. Fucus vesiculosus, *brown alga consumed as a sea vegetable in Europe (photo credit © Fleurence J. 2021)*

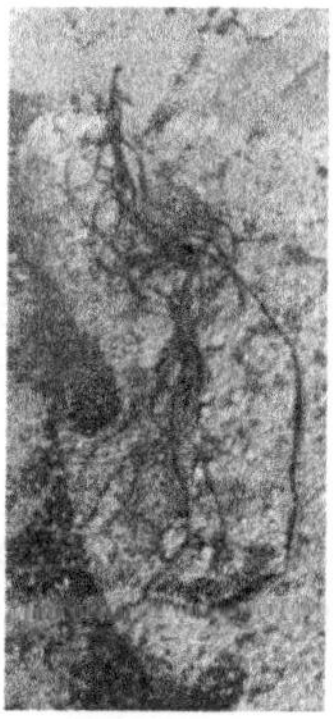

Figure 6.6. Himanthalia elongata *(sea spaghetti), brown alga consumed as a sea vegetable in Europe (photo credit © Fleurence J. 2021)*

Regional names
Dabberlocks
Badderlocks
Mulins
Ribini
Cupognag cloc

Table 6.4. *Regional names in the United Kingdom for the edible alga* Alaria esculenta *(Atlantic Wakame) (see: https://seaweedsolutions.com/fr/alarie-succulente-Wakame-alaria-esculenta/)*

Green algae are also valued as sea vegetables and, like other types of algae, are mainly consumed in Asian countries. In Japan, these algae are often consumed in the form of a mixture of several species (*Ulva* spp., *Enteromorpha* spp., *Monostroma* spp.). This mixture is known as Aonori in Japanese or "green laver" in English (Nisizawa et al. 1987). The commercial name of this mixture may be identical to that given to each component of the product taken separately and may be a source of confusion (Table 6.5). Indeed, the term "green laver" refers both to Aonori, which is a mixture of at least three species, and to algae of the genus *Enteromorpha.*

Species	Trade name of the product
Mixture *Ulva* spp. *Enteromorpha* spp. *Monostroma* spp.	Aonori or "Green laver"
Ulva spp.	Sea lettuce
Enteromorpha spp.	Green laver
Monostroma spp.	Hitoegusa

Table 6.5. *Commercial names in Japan for products made from green algae of the genera* Ulva, Enteromorpha *and* Monostroma *(Nisizawa et al. 1987)*

In France, algae of the genus *Enteromorpha* sp. are used as sea vegetables under the generic name Aonori, which may cause confusion for this product, particularly when exported.

In addition to the species mentioned above, other green algae are also commonly consumed as sea vegetables, particularly algae of the genus *Caulerpa.* The species *Caulerpa lentillifera* is an alga frequently consumed as a sea vegetable in Japan, particularly on the island of Okinawa. In Japan, it is known as Umibudo, which literally means "sea grapes" (Figure 6.7).

This example highlights the cultural difference between Asia and Europe in the perception of algae as food. Among the Caulerpa species, *Caulerpa taxifolia*, better known as Caulerpa, has an extremely unfavorable image in Europe due to its ecological behavior as an invasive alga, particularly on the southern coast of France. However, this species is consumed by local populations in the Philippines as a sea vegetable (Zemke-White and Ohno 1999). *Caulerpa racemosa* and *C. lentillifera,* mentioned above, are also integrated into the diet of Asian countries (Japan, Vietnam, Philippines), the Indian peninsula (Bangladesh) and the South Pacific islands (Zemke-White and Ohno 1999).

Figure 6.7. Caulerpa lentillifera *or Umibudo (photo credit © see: https://commons.wikimedia.org/w/index.php?curid=6913523*

Algae of the genus *Ulva* are green algae commonly consumed in Europe (Figure 6.8). These algae are marketed under various names such as "fisherman's salad" or "sea lettuce". They are mainly available in the form of dried flakes or salted seaweed.

Figure 6.8. Ulva *sp. or sea lettuce (photo credit © Fleurence J. 2021)*

Red algae are also consumed as sea vegetables. Algae of the genus *Porphyra* are particularly popular in Asia and now in the West, thanks to consumer enthusiasm for maki rolls (Figure 5.6). In this type of product, the alga, or Nori, can be considered a sea vegetable or an intermediate food product (Chapter 6, section 6.3). In Japan, two species of red algae are used to make Nori. These are *Porphyra yezoensis* and *P. tenera.* In France, the name "Nori" is reserved for the species *P. umbilicalis.* Several types of food products, consumed as sea vegetables or used as intermediate products, are derived from Nori. The main product is known as "Hoshi-Nori". These are dried Nori sheets (Chapter 5, Figure 5.6).

In addition to Hoshi-Nori, other products such as grilled Nori and seasonal Nori are available to consumers. In addition to these basic products, more elaborate products such as "Yaki-Nori", a type of roasted Nori, or "Tsukudani-Nori", Nori blanched in soy sauce, are also available for consumption as a sea vegetable (Figure 6.9).

Figure 6.9. *Tsukudani-Nori, grilled and blanched Nori seaweed in soy sauce (photo credit © Fleurence J. 2024)*

In France, Nori is mainly sold in flake form and is used more as an ingredient or intermediate food product (Chapter 6, section 6.3). Other red algae are also valued as sea vegetables. In Japan, species of the genus *Gracilaria* are consumed as salad under the name "Ogonori" or "Ogu-Nori" (Ohno and Largo 1998) (Table 6.6).

It is interesting to note that the species *Grateloupia filicina* is also used as an edible seaweed salad under the name "Mukadenori", while algae belonging to the genus *Grateloupia* are not authorized for human consumption in France (Table 6.6).

In Korea, many red algae are also consumed as salads or sea vegetables in soups (Tables 6.6 and 6.7). As in Japan, algae of the genus *Grateloupia* are among the edible algae. The species *Grateloupia turuturu* (Yamada 1941), also known as Grateloupe (Figure 6.10), is used as food in the Far East (da Costa et al. 2021). In Europe, this non-native alga is considered to be proliferant, or even invasive, and is not used as a sea vegetable on the European continent.

Species	Japanese trade name	Product status	Food use	Note
Porphyra blend	Nori	Dried	Maki rolls	*P. tenera* *P. yezoensis*
Grateloupia filicina	Mukadenori	Salted	Salad	–
Eucheuma gelatinae	Kinrisai	Salted	Salad	–
Meristotheca papulosa	Tosakanori	Salted	Salad	–
Gracilaria blend	Ogonori	Salted	Salad	*G. verrucosa* *G. lemaneiformis* *G. bursa-pastoris*

Table 6.6. *Red algae used as sea vegetables or intermediate food products in Japan (Ohno and Largo 1998)*

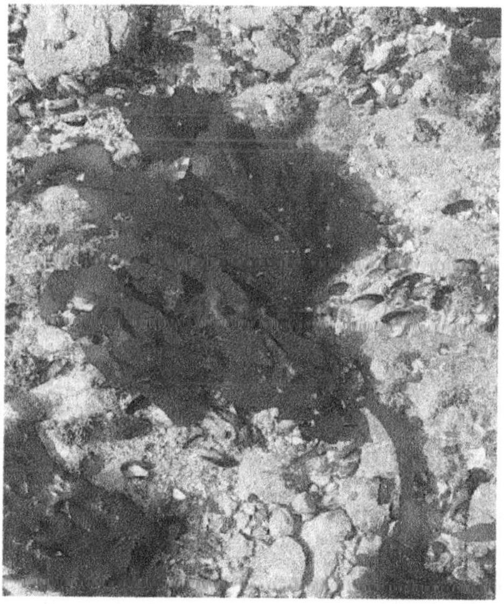

Figure 6.10. Grateloupia turuturu, *red alga consumed in Korea and Japan as a sea vegetable (photo credit © Fleurence J. 2023)*

Species	Korean trade name	Product status	Food use	Note
Porphyra blend	Kim	Fresh or dried	Grilled, maki rolls	*P. tenera* *P. yezoensis*, *P. kuniedae* *P. seriata* *P. suborbiculata*
Grateloupia turuturu	Jinuari	Fresh or dried	Salad	–
Gracilaria verrucosa	Kkoshiraegi	Fresh or dried	Seasonal sea vegetable	–
Chondria crassicaulis	Seoshil	Fresh or dried	Seasonal sea vegetable	–
Gloiopeltis furcata	Bultunggasari	Fresh or dried	Soups, seasonal sea vegetable	–
Gelidium tenax	Chamgasari	Fresh or dried	Soups, seasonal sea vegetable	–
Nemalion vermiculare	Guksunamul	Fresh	Salad	–

Table 6.7. *Red algae used in Korea as sea vegetables or ingredients (Sohn 1998)*

In Europe, and particularly in France, only the species *G. verrucosa* is used as a sea vegetable and is mainly sold as dried flakes. The red alga *Palmaria palmata*, formerly *Rhodymania palmata*, is a species consumed as a sea vegetable mainly in Europe and the West. This species is well known under various common names such as "*Rhodoménie palmée*", "Dulse" (France), "Dillisk" (United Kingdom) and Creathnach (Ireland). In Ireland and France, this alga is eaten in soups or as a vegetable in stews (Mouritsen et al. 2013) (personal communication, Joël Fleurence 2024). In France, Dulse is packaged in the form of dried flakes or mixed with other algae (Ulva, Porphyra) to make "seaweed tartare". It can therefore be used as an ingredient, an intermediate food product or a sea vegetable. Some algae commonly used in human food have recently had their species names changed (Table 6.8). This change is not yet noticeable in commercial products but should gradually become established.

Old name	New name
Laminaria japonica (Kombu)	*Saccharina japonica* (Kombu)
Laminaria saccharina (Royal Kombu)	*Saccharina latissima* (Royal Kombu)
Porphyra tenera (Nori)	*Pyropia tenera* (Nori)
Gracilaria verrucosa (Ogonori)	*Gracilaria gracilis* (Ogonori)
Enteromporpha compressa (Aonori)	*Ulva compressa* (Aonori)
Enteromorpha intestinalis (Aonori)	*Ulva intestinalis* (Aonori)

Table 6.8. *Algae used as sea vegetables whose scientific name has changed*

6.3. Ingredients and intermediate food products

The use of algae as sea vegetables, ingredients or intermediate food products meets with a definition that may vary depending on the actors in the sector (processors, consumers, government agencies).

The use of algae in traditional preparations such as Dashi soups or as an accompaniment to certain dishes is considered to be a "vegetable" use. In Japan, algae are also considered a sea vegetable, regardless of how they are incorporated into a dish (e.g. maki rolls) (Figure 6.11). In Europe, however, a distinction is made between vegetables, ingredients and intermediate food products.

Figure 6.11. *Maki rolls containing a strip of Nori (sea vegetable, ingredient or intermediate food product) (photo credit © Barbaroux O. 2018)*

Industrial and regulatory practice distinguishes between the concept of an ingredient and that of an intermediate food product. An intermediate food product is a product that has already undergone partial processing, i.e. it has undergone a transformation process, which excludes preservation processes such as drying, salting or freezing. Dried and flaked algae are therefore ingredients in the strict sense of the term. On the other hand, algae that have undergone pressure cooking, such as appertization, are considered to be products that have already undergone processing and are classified as intermediate food products, if they are used in the composition of a processed product. However, French regulations (Chapter 7, section 7.3.1) only take into account the concept of "sea vegetables" or "ingredients" in their authorization of algae for human consumption (Fleurence 1991).

Many products incorporating algae as ingredients have appeared since the late 1980s. The best known is "seaweed bread". Among these, seaweed baguettes have enjoyed considerable commercial success. This traditional and typically French product is made with seaweed flakes, usually Ulva, Dulse and Nori (Figure 6.12). It is a purely artisanal, typically regional and seasonal product. It is mainly found in the departments of Brittany primarily during the summer period.

Figure 6.12. *Seaweed baguette (Dulse, Ulva, Nori) (photo credit © Barbaroux O. 2018)*

Other products incorporating algae as ingredients are also available. This is the case for canned fish (sardines and mackerel with algae) (Figure 6.13). The algae used may be sea beans (*Himanthalia elongata*), sea lettuce (*Ulva* sp.) or Wakame (*Undaria pinnatifida*).

Figure 6.13. *Canned sardines with seaweed (Wakame) (photo credit © Fleurence J. 2024)*

Algae are also widely used as an ingredient in various culinary preparations such as algae tartares (Figure 6.14). In this type of product, they are the main ingredient although they are usually present in a mixture of three species (sea lettuce, Dulse, Nori and/or sea beans). In France, they are also found as ingredients in traditional products such as fish rillettes, mainly made from sardines, mackerel or tuna.

Algae can also be incorporated as an ingredient in pasta. Dulse pasta (Figure 6.15) and sea lettuce pasta are available from manufacturers in the sector. During product development, some manufacturers may also use a mixture of algae (Fucus, sea lettuce) (Figure 6.16). In general, algae account for 5–15% of the final product (Table 6.9).

Figure 6.14. *Spreadable algae tartare made from sea lettuce (*Ulva *sp.), Dulse (*Palmaria palmata*) and Wakame (*Undaria pinnatifida*) (photo credit © Fleurence J. 2024)*

Figure 6.15. *Pasta made from Dulse (*Palmaria palmata*)*
(photo credit © Bord à Bord 2017)

Figure 6.16. *Example of pasta incorporating algae (*Ulva *sp. and* Fucus *sp.) during the production of the final product (photo credit © Fleurence J. 2024)*

The incorporation of algae as a food ingredient in pasta has also been studied to evaluate the chemical, functional and structural properties of the final product (Prabhasankar et al. 2009).

These studies have mainly focused on the use of Wakame as an ingredient. The incorporation of Wakame at a level of 20% gives good results in terms of organoleptic quality, but above all allows the best properties to be obtained in terms of nutritional value (amino acid composition, polyunsaturated fatty acids) and functional properties of the product. However, adding 10% gives the best score in terms of sensory perception of the final product. The addition of Wakame also increases the nutritional value of the product through the contribution of biofunctional compounds such as fucoxanthin and fucosterol, which are specifically provided by the algae. Furthermore, the cooking process has no impact on fucoxanthin, making this product a dietary source of this pigment for human consumption (Prabhasankar et al. 2009).

Main ingredients	**%**
Durum wheat	95
Algae (Fucus, sea lettuce)	5
Nutritional value	**Per 100 g of dry product**
Dietary fiber	3.9 g
Carbohydrates	60.5 g
Fat	0.9 g
Protein	11 g
Salt	0.179 g
Energy	300 kcal

Table 6.9. *Example of ingredient composition of algae-based pasta*

Appertized algae have also been used as an intermediate food product or semi-finished product for the development of finished products incorporating algae, particularly Laminaria.

At the end of the 1980s, when the "edible algae" industry was booming in France, projects emerged for products such as fish cakes with algae, particularly Laminaria (Figure 6.17). Unfortunately, due to a lack of sufficient market demand at the time and given the regulatory constraints applied to algae, particularly Laminaria (Chapter 7, section 7.3.1), this type of product was never marketed.

Lactofermented algae (based on sea beans) was also developed in the late 1980s as an intermediate food product. This product, described as "seaweed sauerkraut", was used to make "ocean salads" by a large industrial caterer (personal communication, Joël Fleurence). A range of products based on fermented soybeans also incorporated lactofermented algae as an intermediate food product.

Figure 6.17. *Prototype food product incorporating algae (*Laminaria*) as an ingredient (photo credit © Barbaroux O. 2018)*

More recently, algae have been used as ingredients in a product traditionally popular with young consumers: burgers. A Dutch company, The Seaweed Company, has developed a new burger called the "Seameat Burger®" (Figure 6.18). This burger is made from 75% beef and 25% a mixture of algae, the species composition of which is not disclosed by the manufacturer. According to information provided by the manufacturer on its public website (Table 6.10), the ecological impact of producing this burger appears to be lower than that of a traditional burger.

Figure 6.18. *Seameat Burger®, marketed by The Seaweed Company (photo credit © Kraan S. and The Seaweed Company 2024)*

The production of one kilogram of Seameat Burger results in a 25% reduction in CO_2 emissions compared to the production of one kilogram of a traditional burger (100% beef). Similarly, a significant saving in freshwater consumption is observed in the production of one kilogram of algae-based burgers. This reduction in water use is approximately 24% compared to the process used to produce burgers made from 100% beef (Table 6.10).

In this context, the use of algae as a partial substitute for beef protein appears to be a new alternative for the use of algae as ingredients or intermediate food products.

	Classic burger (100% beef)	Seameat Burger (75% beef, 25% algae)
Greenhouse gas production (kg CO_2 equivalent)	33.2	25.1
Freshwater consumption (L)	6,850	5,210
Water pollution (g PO4 equivalent)	2	1.5
Land area used (m^2)	89	66.8

Table 6.10. *Ecological balance sheet of the production of 1 kg of Seameat Burger versus a traditional burger (see: https://www.theseaweedcompany.com/our-products/seameat)*

The use of algae in combination with animal proteins other than beef protein has also been tested in a typical seafood burger. In this product, the main source of animal protein comes from fish belonging to the Scombridae family. In addition to being rich in protein, fish from this family are characterized by high levels of polyunsaturated fatty acids. In this nutritional context and with the aim of creating a functional food, a seafood burger incorporating mackerel meat and algae was the subject of a specific study (Cardoso et al. 2024). The aim of this new product is to provide consumers with a food source enriched in polyunsaturated fatty acids, such as docosahexaenoic acid (DHA). The burger contains Spanish mackerel (*Scomber colias*) and brown alga *Saccorhiza polyschides (Laminaria bulbosa*), commonly known as bulbous kelp (Figure 6.19).

Figure 6.19. Saccorhiza polyschides (Laminaria bulbosa) (*photo credit © see: https://fr.wikipedia.org/wiki/Saccorhiza_polyschides#/media/File:Alga_Golfo_Saccorhiza_polyschides2.jpg)*

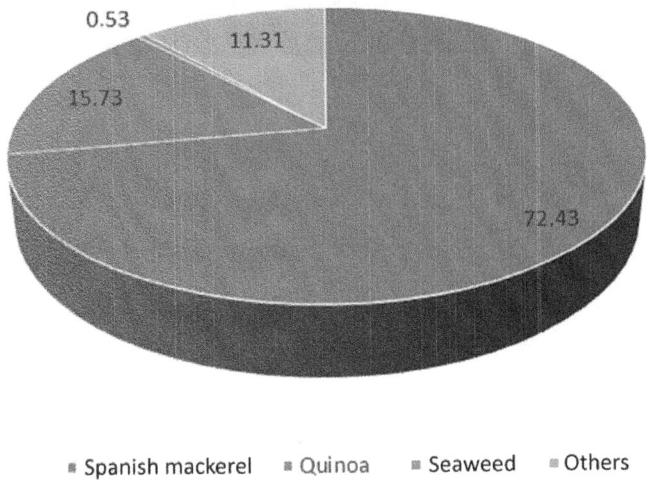

Figure 6.20. *Composition of the fish and algae burger (expressed as a % of the final product) (Cardoso et al. 2024)*

In this food formulation, algae are not very prevalent (0.53% of the final product) compared to mackerel and quinoa, which are the main ingredients of this burger (Figure 6.20). As an ingredient, even if minor, it reinforces the image of a seafood burger already conveyed by the mackerel meat. In this type of product, algae cannot be considered an intermediate food product. This product is very different from the Seameat Burger, which contains up to 25% algae in the final composition of the burger. Furthermore, the docosahexaenoic acid content, which is the main nutritional benefit justifying the production of this algae burger, is due to the presence of mackerel meat in the final product (Figure 6.21).

Other burgers intended for vegan consumers also use algae in the final composition of the product. The algae concerned are listed under the name "Kelp", which remains a common name applied to many species of brown algae (*Macrocystis* spp., *Saccharina* ssp., *Alaria* spp., etc.). The Sea-Veggie Burger is characterized by its entirely plant-based composition (Table 6.11). The list of ingredients provided by the company responsible for manufacturing this product does not include the relative contents of each compound. Similarly, the nature of the brown algae used is not disclosed by the processor. The Kelp Burger is also a product intended for vegetarian or vegan consumers. Like the Sea-Veggie Burger, it uses brown algae, but it differs from the latter in particular through the use of button mushrooms, konjac, quinoa and, above all, amino acids extracted from coconut (Table 6.11). The Kelp Burger also contains agar. This algae-based colloid reinforces the "seaweed" image associated with this product. Finally, the Kelp Burger does not contain sesame oil or seeds. This avoids the use of sesame, which is a well-known allergenic ingredient.

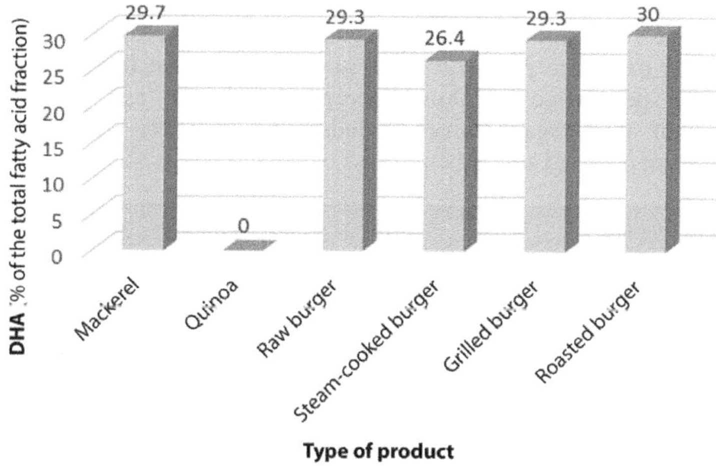

Figure 6.21. *Docosahexaenoic acid (DHA) content in the main burger components and the different burgers before and after cooking (Cardoso et al. 2024)*

Sea-Veggie Burger	Kelp Burger
Brown algae (kelp)	Brown seaweed (kelp)
Pea protein	Pea protein
Green chickpeas	Quinoa
Brown rice	Button mushrooms
Ginger sesame oil	Virgin olive oil
Pea starch	Potato starch
Chickpea flour	Black beans
Oat fiber	Konjac
Sesame seeds	Tomato powder
Garlic	Spices
Nutritional yeast	Nutritional yeast
Kosher salt (rock salt)	Agar
–	Coconut amino acids

Table 6.11. *Relative composition of the Sea-Veggie Burger (Atlantic Sea Farms) and the Kelp Burger (Akua) (see: https://atlanticseafarms.com/product/ginger-sesame-sea-veggie-burgers/ and https://akua.co/pages/kelpburgervsfakemeat)*

Brown algae are not the only species that can potentially be used to make vegan burgers. Recently, work on the development of a burger incorporating the red alga *Palmaria palmata* or Dulse was published internationally (Fernandes et al. 2023). The study also looked at the acceptance of this type of product by a sample of consumers. The final product is mainly composed of chickpeas (*Cicer arietinum*), common vetch or square peas (*Lathyrus sativus*) and Dulse (*Palmaria palmata*). However, the percentage of Dulse incorporated is not specified. The final composition of the Vegan Burger includes approximately 10 ingredients of exclusively plant origin (Table 6.12).

Ingredients	%
Chickpea/Lathyrus mixture (50/50)	Not provided
Palmaria palmata	
Grated carrot	
Onion	
Oat flakes	
Lupin flour	
Apple cider vinegar	
Olive oil	
Herb and spice mix	

Table 6.12. *Ingredient composition of the Vegan Burger made from Dulse (Fernandes et al. 2023)*

The Dulse-based Vegan Burger is characterized by a high protein content when calculated on a dry weight basis. The same is true for carbohydrates and particularly for fiber, whose nutritional benefits are now well known. However, the product's sugar content is particularly low compared to its fiber content, which is a nutritional advantage for this Dulse-based burger (Figure 6.22).

The lipid fraction of the product is characterized by an interesting distribution between saturated and unsaturated fatty acids (Figure 6.23). It is particularly notable for its high content of unsaturated fatty acids, especially polyunsaturated fatty acids (58.47% of the total fatty acid fraction). Linoleic acid (C18:2) is the most abundant polyunsaturated acid, accounting for nearly 53.65% of the total fatty acid fraction. Among the monounsaturated acids, oleic acid (C18:1) is the most abundant monounsaturated fatty acid. It accounts for 21% of the total fatty acid fraction. The high level of unsaturated fatty acids and the predominance of certain fatty acids, such as oleic acid, are probably due to the incorporation of olive oil during the production of the final product.

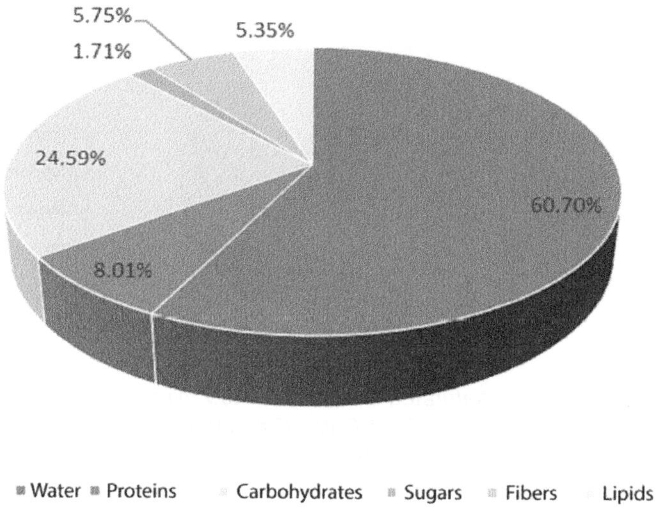

Figure 6.22. *Composition of the vegan burger made from Dulse (*Palmaria palmata*) (expressed as a % of wet weight) (Fernandes et al. 2023)*

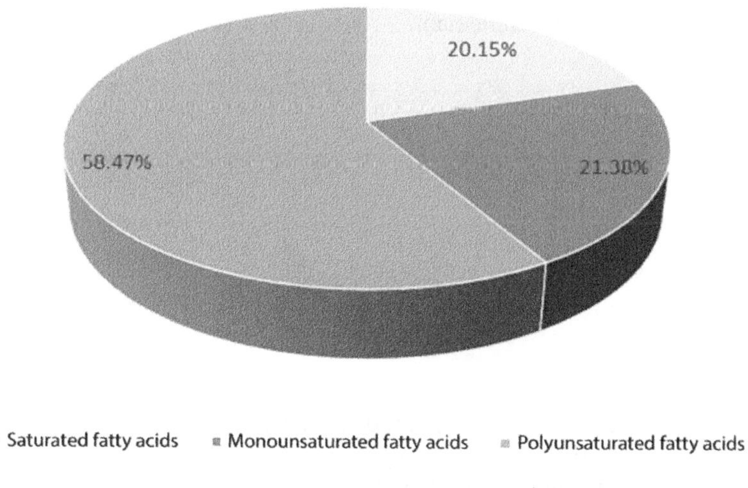

Figure 6.23. *Distribution of saturated and unsaturated fatty acid fractions in the Dulse-based vegan burger (*Palmaria palmata*) (expressed as a % of the total fraction of fatty acids) (Fernandes et al. 2023)*

In terms of mineral composition, the Dulse-based Vegan Burger is characterized by a high content of phosphorus, calcium and magnesium (Figure 6.24).

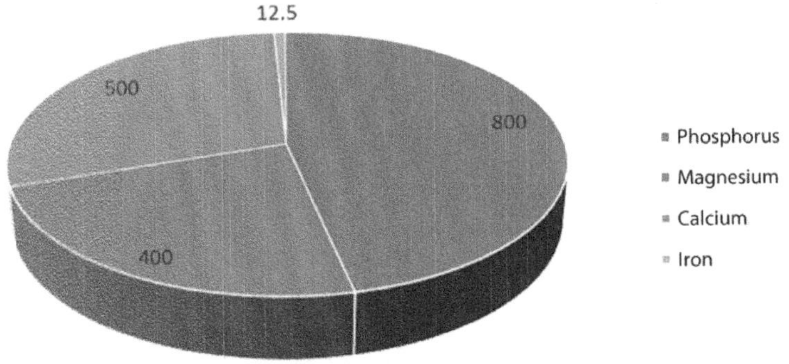

Figure 6.24. *Distribution of saturated and unsaturated minerals of nutritional interest in the Vegan Burger made from Dulse (*Palmaria palmata*) (expressed as mg/kg of fresh matter) (Fernandes et al. 2023)*

This type of product therefore meets the nutritional claims for phosphorus and magnesium as a food source as defined by European regulations (Fernandes et al. 2023)[1]. Furthermore, the acceptability of this type of product by a sample of consumers aged 16–64 was generally satisfactory. 65% of consumers tested found the product pleasant or very pleasant (Figure 6.25). 33% found it somewhat pleasant, and only 2% rated it as very unpleasant.

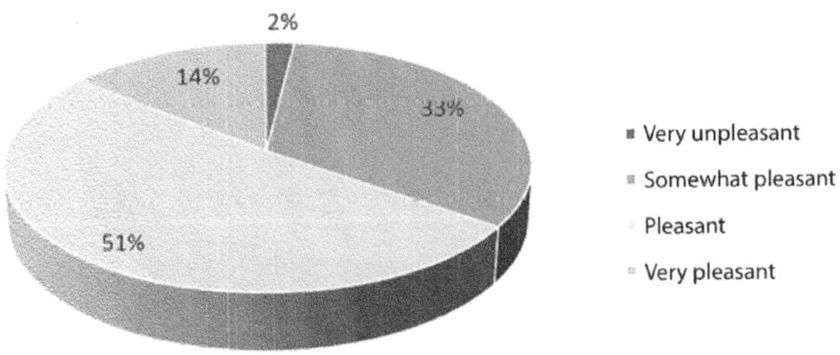

Figure 6.25. *Acceptability test of the Vegan Burger made from Dulse (*Palmaria palmata*) by a sample of consumers (89 people aged 16–64) (Fernandes et al. 2023)*

1. See: https://eur-lex.europa.eu/LexUriServ/LexUriServ.do?uri=OJ:L:2011:304:0018:0063:en:PDF.

Following this initial assessment, the study focused on the sample members' intention to purchase this type of product. Nearly 54% of the consumers selected said they would be willing to buy this type of product if it became commercially available (Figure 6.26). However, 4.5% of the group said they were extremely opposed to such a purchase. Nevertheless, this latest study, although limited in terms of the group size, shows that consumers are open to new flavors and new products made from algae.

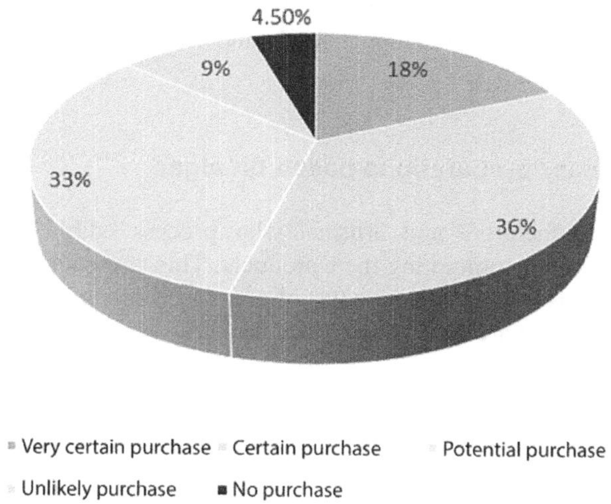

Figure 6.26. *Purchase intention test for the Dulse-based Vegan Burger (*Palmaria palmata*) on a consumer panel (89 people aged 16–64) (Fernandes et al. 2023)*

The integration of algae as an intermediate food product or ingredient in well-known products such as hamburgers is an approach that could promote algae consumption among Western consumers, particularly younger people, who are very fond of this type of product.

6.4. Dietary supplements

The use of algae as dietary supplements is very widespread. The products available come in the form of liquid extracts, usually packaged in drinkable ampules. Algae can also be packaged in powder form, available in capsules or tablets. These products are not subject to the regulatory constraints imposed on algae consumed as an occasional vegetable or food ingredient and are governed solely by food supplement regulations.

Drinkable ampules mainly contain laminaria extracts marketed for their high trace element content, according to the marketing claims. Some liquid algae extracts are also marketed for use in energy drinks. The species used in the production of this type of extract are not always specified, and only the term "seaweed extract" is often mentioned. *Fucus serratus*, an alga authorized by the pharmacopoeia, is used in France to produce capsules sold mainly in pharmacies in the nutrition section.

Apart from the products currently often sold online, the use of seaweed as a food supplement or additive is also the subject of extensive research. These studies are mainly aimed at developing functional foods and will be discussed in more detail later (Chapter 8, section 8.3).

6.5. Some culinary preparations based on algae

Industrial manufacturers and artisans who process edible algae now offer culinary preparations incorporating their products. This approach is becoming more widespread and aims to normalize the use of algae in cooking by combining it with traditional products that are well known to consumers.

This is particularly the case with "sea lettuce purée" and "Dulse quiche" (Box 6.1).

Sea lettuce purée (serves 4) (Bord à Bord 2017)

Ingredients

500 g pureed potatoes

25 g sea lettuce

75 g butter

10 cl milk

10 cl cream

Method

Peel the potatoes, cut into large cubes and cook in salted water.

Finely chop the sea lettuce.

Once cooked, mash the potatoes and gradually add the butter, milk and cream, stirring until you achieve the desired consistency.

Finally, add the sea lettuce.

The purée can also be used to make a fish Parmentier.

Figure 6.27. *Sea lettuce purée (photo credit © Bord à Bord commercial* 2017)

Dulse quiche (Bord à Bord 2017)

Ingredients

4 medium eggs

40 g Dulse

15 cl cream

10 cl milk

2 slices of smoked bacon, grated Gruyère or Comté cheese

Puff pastry or shortcrust pastry

Method

Finely chop the Dulse.

Whisk the whole eggs in a bowl and add the cream, milk and Dulse.

In a frying pan, fry the smoked bacon cut into slices.

Preheat the oven to 180°C.

Roll out the pastry and line a quiche dish, then place the bacon on the tart base.

Add the grated cheese and bake for 25–30 minutes.

Figure 6.28. *Dulse quiche (photo credit © Bord à Bord commercial 2017)*

Box 6.1. *Examples of culinary preparations using algae*

6.6. Iodine intake from algae consumption

Iodine is an essential nutrient and deficiency can lead to thyroid disorders. The deficiency threshold for adults is estimated to be a dietary intake of less than 150 µg/day. The European Food Safety Authority (EFSA) has established a safe upper intake level (UL). This upper limit for dietary iodine intake is set at 600 µg/day (Dopter et al. 2021). The main sources of iodine are marine resources and iodized table salt (Table 6.13).

Food product	Iodine content (µg/100 g)
Periwinkle	570
Langoustine	394
Haddock	260
Breton Kombu (*Laminaria digitata*)	486
Royal Kombu (*Laminaria saccharina*)	341
Ogonori (*Gracilaria verrucosa*)	494
Egg yolk	175
Sheep's milk cheese	124
Enriched table salt	1,860

Table 6.13. *Examples of iodine content in some products of marine origin, including algae (Dopter et al. 2021)*

Algae, like other marine food resources, are particularly rich in iodine. However, the content of this nutrient varies between species and, for a given species, between seasons. Interspecific variability is particularly high. The red alga *Porphyra* sp., consumed under the name Nori, has an average iodine content of 51 µg/g (DM), compared with an average content of 4,850 µg/g (DM) for *Laminaria digitata* (Breton Kombu) (Dopter et al. 2021). In some cases, the iodine content of *L. digitata* can even reach 10,000 mg/kg of dry matter.

Intra-species variability in iodine concentration can also be very high. For the alga *L. digitata*, a variation of a factor of 5.5 is observed between the minimum and maximum iodine levels (Figure 6.29).

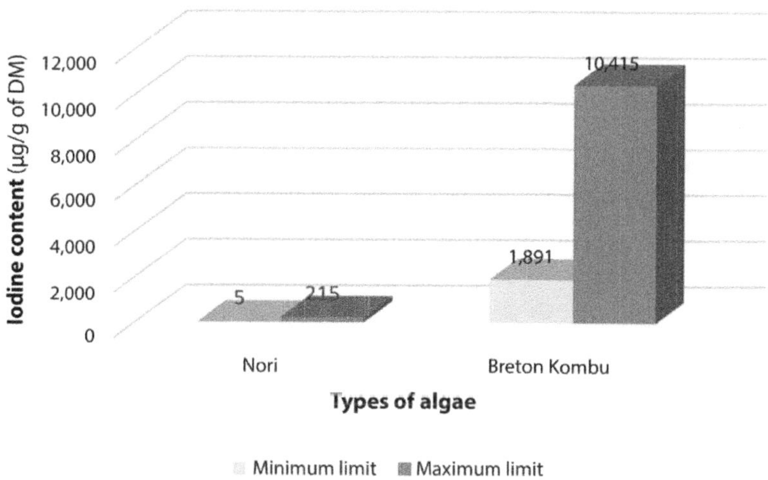

Figure 6.29. *Variation in the iodine content of algae used as sea vegetables, intermediate food products or ingredients (Dopter et al. 2021)*

Most algae offered in raw form, i.e. dried, in a 7 g serving, have high iodine levels that exceed the food safety limit established in Europe (Table 6.14).

A more nuanced result is described for two species of red algae, namely Pioca (*Chondrus crispus*) and Nori (*Porphyra* sp.). The iodine content reported for these two species cover 85% and 96% of the minimum iodine intake (150 µg) (Figure 6.30). However, consumption of these algae in the presence of other food sources could lead to the food safety limit (600 µg/day) being reached.

Algae	Iodine content (mg/7g of dry product)
Blanched Royal Kombu (*Laminaria saccharina*)	3.7
Wakame (*Undaria pinnatifida*)	2.6
Atlantic Wakame (*Alaria esculenta*)	2.7
Sea bean (*Himanthalia elongata*)	0.5
Sea lettuce (*Ulva lactuca*)	1.2
Nori (*Porphyra* sp.)	0.4
Dulse (*Palmaria palmata*)	0.1

Table 6.14. *Contribution of a small amount of dried algae (7 g) to the recommended daily intake of iodine (Marfaing 2017)*

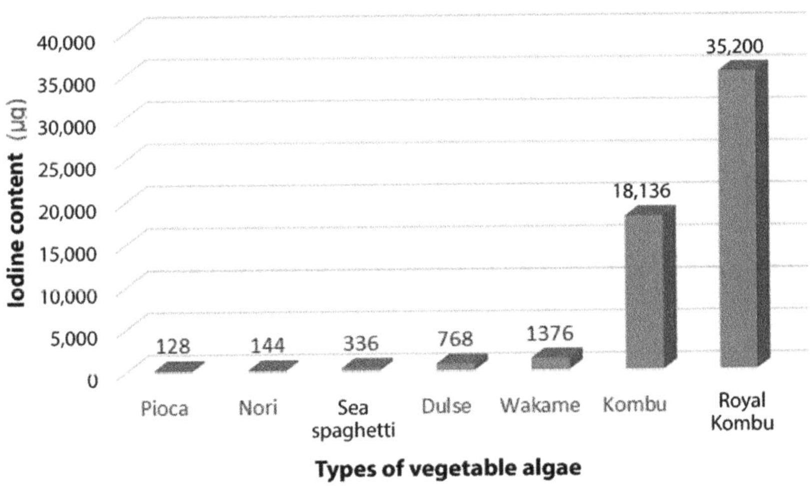

Figure 6.30. *Iodine content (µg) of a serving of vegetable alga (8 g) (Aarkre et al. 2021)*

However, this finding must be viewed in context, as the levels reported mainly concern raw material as it comes out of the water and has been stabilized by drying.

On the other hand, certain processing steps have proven particularly effective as a method of iodine removal. This is particularly the case with boiling (blanching) or steaming, which removes up to 58% of the iodine contained in most algae (Dopter et al. 2021). Analyses carried out on appertized Laminaria (*Laminaria digitata*) have shown that the industrial process based on soaking the algae in fresh water, followed by drying and subsequent rehydration, then appertization, resulted in the removal of nearly 80% of the iodine initially measured in the raw material (personal communication, Joël Fleurence). In *Laminaria japonica* (Kombu), boiling for 15 minutes removes 99% of the iodine contained in the alga. However, this leaching of iodine under the same conditions is less effective for algae of the genus *Sargassum* (– 40% iodine removal) (Zava and Zava 2011).

Furthermore, incorporating algae as an ingredient in food products can also reduce the risk of excessive iodine consumption, as the alga is "diluted" in the finished product.

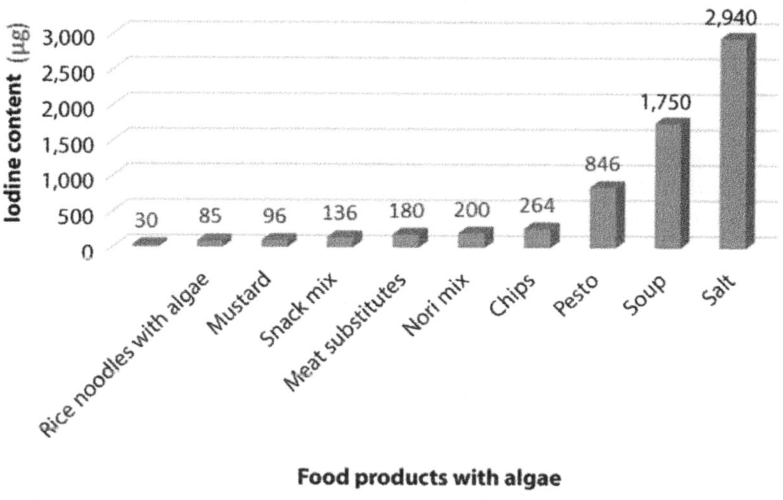

Figure 6.31. *Iodine content (µg) in a serving of a food product containing algae (8 g) (Aakre et al. 2021)*

The bioavailability of algal iodine is an important parameter for assessing the toxicity risk associated with this nutrient. This criterion makes it possible to evaluate the accessibility of iodine to the digestive system and its absorption by the latter. This bioavailability can be assessed using in vitro approaches such as solubility, dialyzability, transport and absorption of iodine via intestinal cells (Caco-2, HT29-MTX cells) (Domínguez-González et al. 2017). After gastrointestinal digestion of algae, the percentage of solubilized iodine (bioaccessibility) differs depending on the species (Figure 6.32). This bioaccessibility is high with Kombu (82%) and lower with Wakame (49%).

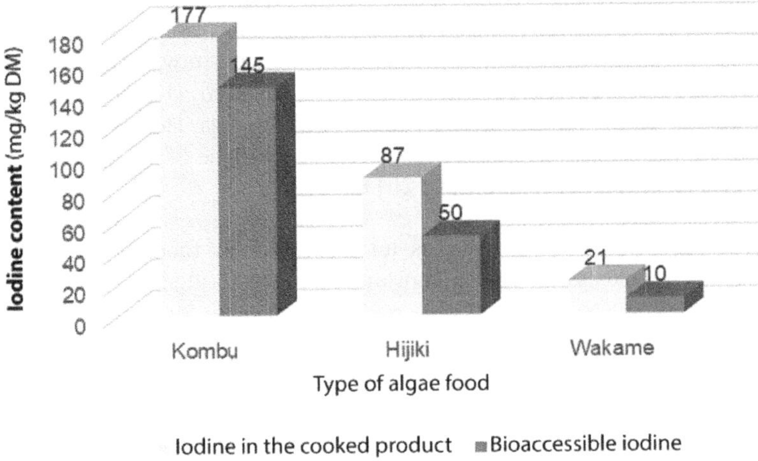

Figure 6.32. *Iodine content of various cooked brown algae products (Kombu, Hijiki, Wakame) and in vitro bioaccessibility of iodine after gastrointestinal digestion (Domínguez-González et al. 2017)*

The absorption and transport of bioaccessible iodine by Caco-2 and HT29-MTX cells have also been studied. The results obtained show that only a small proportion of the available iodine is absorbed and transported in vitro by intestinal cell cultures. In the case of the brown alga Hijiki (*Hizikia fusiformis*), maximum iodine absorption is capped at 6% (Figure 6.33).

Similarly, for Kombu (*Laminaria japonica/Saccharina japonica*), a very small proportion of the available iodine is transported by the Caco-2, HT29-MTX system (Figure 6.34). Somewhat paradoxically, the highest transport rate of accessible iodine is recorded with Wakame (7%) (Domínguez-González et al. 2017).

These results show that the in vitro availability of iodine from brown algae rich in this nutrient is relatively low. They also reveal that absorption by intestinal cells differs depending on the algal origin of this iodine. This finding suggests that there are different chemical forms of iodine depending on the species, which probably explains the observed differences in the behavior of digestive cell systems with regard to this nutrient.

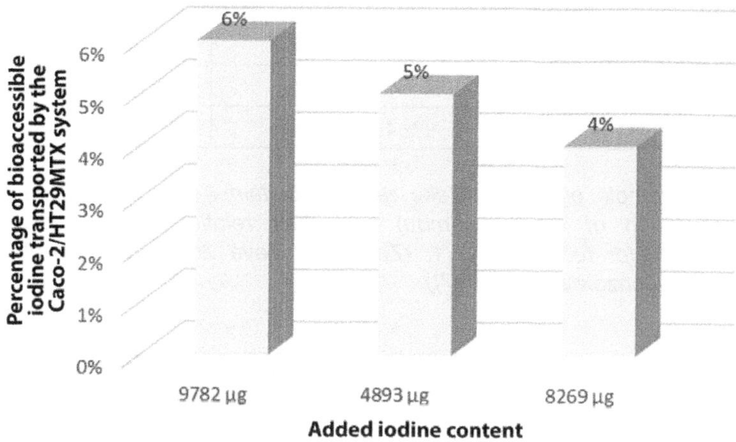

Figure 6.33. *Fraction of bioaccessible iodine content (expressed in µg) of Hijiki alga transported and absorbed in vitro by the Caco-2, HT29-MTX system (Domínguez-González et al. 2017)*

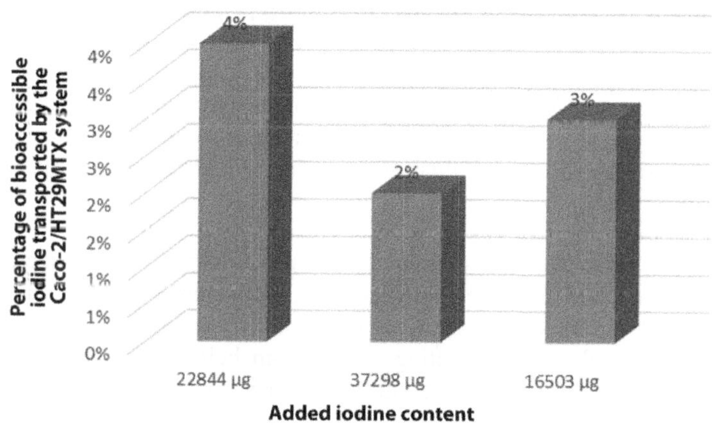

Figure 6.34. *Fraction of bioaccessible iodine content (expressed in µg) of Kombu alga transported and absorbed in vitro by the Caco-2, HT29-MTX system (Domínguez-González et al. 2017)*

Asian consumers of algae ingest amounts of iodine well above the food safety limit recommended by EFSA (Table 6.15). Unassimilated iodine is excreted in the urine (Table 6.16) (Yeh et al. 2014).

Population	Iodine intake (µg/day)	Relative to the LSS
Japanese (1)	1,000–3,000	+1.6–5 times
Korean (2)	4,086	+6.8 times
Chinese (Taiwan) (3)	> 4,800	+8 times

Table 6.15. *Example of iodine intake rates in certain Asian populations following daily consumption of algae (Kombu) and their relationship to the European maximum levels for food safety (1: (Zava and Zava 2011). 2: (Marfaing 2017). 3: (Dominguez-Gonzalez et al. 2017))*

	Iodine excreted in urine (µg/L)
Men > 19 years	102.5
Women > 19 years	97.7
Men > 65 years old	99.3
Women > 65 years old	77.4

Table 6.16. *Elimination of iodine from algae in urine (Yeh et al. 2014)*

The exposure of Asian populations to high iodine intake linked to daily consumption of algae is undeniable. The imperfect assimilation of this nutrient by the digestive system and its elimination via urine would partially limit the toxicity inherent in high doses.

However, other explanations may reinforce this finding. In Asian societies, algae are commonly cooked with broccoli, cabbage, bok choy (Chinese cabbage) and soybeans. Certain molecules, such as the isothiocyanates found in cruciferous vegetables, are capable of preventing iodine from being absorbed by the thyroid gland. Similarly, the isoflavones present in soybeans can inhibit the incorporation of iodine into the thyroid hormone (Zava and Zava 2011).

Regardless of these explanations, it remains that human populations may also have adapted more specifically to certain diets. The consumption of algae (Hijiki, Kombu, Arame) by Japanese populations dates back nearly 10,000 years (Fleurence 2016, 2018). In this context, an adaptation of populations to the consumption of products with a high iodine content also remains an explanation that is not without interest.

In Western countries, a few cases of thyroid disorders have been reported following the consumption of foods containing algae. In Australia, in the late 2000s, eight people aged 29–47 suffered from this type of disorder after consuming a soy drink containing Kombu with an iodine concentration of 25,000 µg/L. In Europe, a case of hyperthyroidism was also reported in a 70-year-old woman who consumed a Fucus-based dietary supplement that provided her with a daily dose of 336 µg of iodine (Dopter et al. 2021). These incidents prompted warnings from government agencies responsible for food safety.

In this sometimes highly controversial context surrounding algal iodine, what can Western consumers do? They can limit their consumption of algae to occasional consumption. This type of dietary behavior is already recommended and practiced for other foods and beverages that are part of the cultural heritage and gastronomic habits. If consumed more regularly, or even daily, consumers should focus on processed products (boiled, pickled) whose processing ensures that most of the iodine is removed. When marketed, these products must comply with the iodine content recommended by the regulatory text on the authorization of algae in human food (Fleurence 2018).

6.7. Phycocolloids (additives)

Algae are also valued in human nutrition as a source of additives with thickening and even gelling properties. In the latter case, the substances responsible for these technological properties are known as phycocolloids. Algae synthesize various polysaccharides (Chapter 4, section 4.2) that have a strong affinity for water. This affinity allows these polysaccharides to generate colloidal solutions when in the presence of water, hence their name "phycocolloids". Approximately 43% of global algae production is used for the extraction of algal colloids (Perez 1997).

There are three main groups of phycocolloids used as additives: alginates produced by brown algae, and agars and carrageenans derived from red algae (Chapter 4, section 4.2).

Alginates are the ionized form of alginic acid. This acid is a molecule that is initially insoluble in water but highly hydrophilic, as it can absorb up to 140 times its own volume in this liquid. In the form of salts with monovalent ions (Na^+, K^+, NH_4^+), alginate becomes soluble in water and develops a strong thickening power, leading to an increase in the viscosity of the medium. In the presence of divalent ions such as Ca^{2+}, the guluronic acid sequences of several molecules combine to form an "egg-box-like" structure (Figure 6.35) in which the calcium ions are trapped. This molecular geometric structure (Figure 6.35) leads to the formation of hard, irreversible gels that are resistant to high temperatures (400°C) when cold.

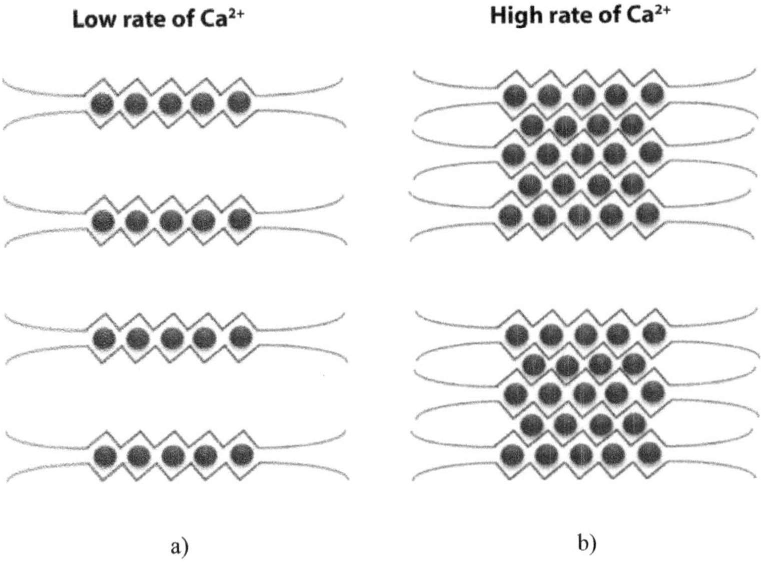

Figure 6.35. *"Egg-box" structure of calcium alginates. a) Low calcium content; b) high calcium content*

Alginates have thickening or gelling properties (Table 6.17) and are listed in the nomenclature of additives under the names E400 to E405. Agars are used for their gelling properties and appear under the name E406 as food additives. Carrageenans are used for their gelling or thickening properties. These compounds are divided into three main families: kappa (κ), lambda (λ) and iota (ι) carrageenans. Kappa and ι carrageenans have gelling properties, while λ carrageenans have thickening properties. Regardless of the family, carrageenans are all listed as E407 in the food additive directory.

Semi-purified carrageenans are also being developed. This type of product is not generally intended for food applications and is marketed under the name "alkaline carrageen floor" (ACF). The process involves treating the biomass with a high concentration of potassium (9% weight/volume) at a temperature of 60°C. This process fixes the carrageenans in the cell wall, and the algae are then dried before being ground into powder. The final product has rheological properties similar to those of purified carrageenans (Perez 1997).

	Additive nomenclature	Rheological properties
Alginic acid	E400	Thickeners or gelling agents
Na alginate	E401	
K alginate	E402	
NH4+ alginate	E403	
Ca alginate	E404	
Propylene glycol alginate	E405	
Agar	E406	Gelling agents
κ-carrageenans	E407	Gelling agents
ι-carrageenans		Gelling agents
λ-carrageenans		Thickeners

Table 6.17. *Rheological properties and nomenclature of additives derived from algae (Fleurence 2018)*

Information is not clear on the global production of phycocolloids (alginates, carrageenans, agar), whereas the tonnages of alginophytes, carrageenophytes and agarophytes harvested for this type of production are generally well known. They are often the subject of publications, whether by the FAO or other sources, as shown in Table 6.18 (Zemke-White and Ohno 1999).

Type of algae	World production
Alginophytes	826,178 tons
Carrageenophytes	81,858 tons
Agarophytes	108,229 tons

Table 6.18. *World production of different types of algae used for the production of alginate, agar and carrageenans (Zemke-White and Ohno 1999)*

Alginophytes are mainly represented by brown algae of the genera *Laminaria* (*Saccharina*) and *Macrocystis* (approximately 90% of production). Carrageenans are obtained from red algae belonging to the genera *Kappaphycus*, *Eucheuma*, *Chondrus* and *Mastocarpus* (95% of production). Finally, agarophytes are mainly algae belonging to the genera *Gelidium* and *Gracilaria* (Table 6.19) (Glicksman 1987). These algae account for almost all of the world's agar production. Many phycocolloid-producing species are cultivated, such as *Laminaria japonica* and *Kappaphycus alvarezii*.

Despite patchy data on global phycocolloid production, some information is available, mainly from the FAO. Global alginic acid production is estimated at between 30,000 and 33,000 tons. This activity is mainly divided between Asia and Europe (Figure 6.36).

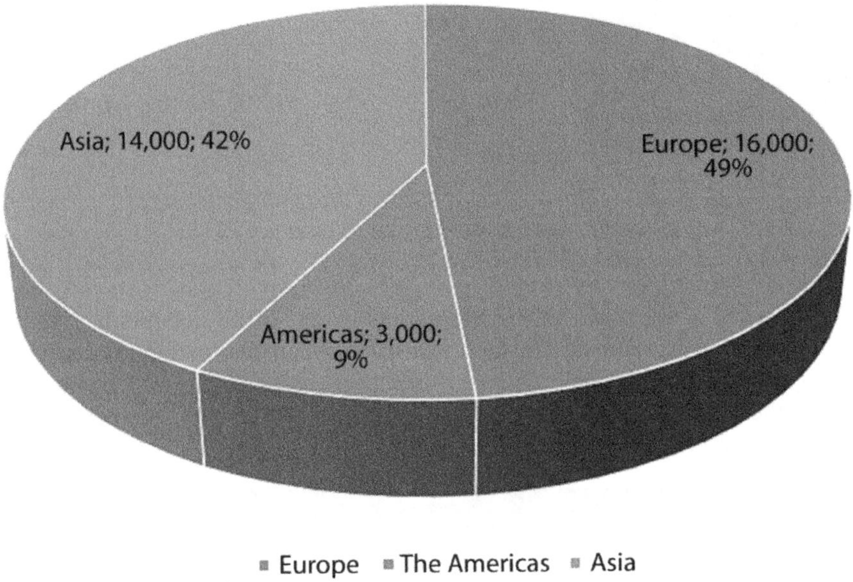

Figure 6.36. *Geographic distribution of global alginic acid production (expressed in tons and %) (see: https://www.fao.org/4/y4765e/y4765e08.htm)*

In comparison, agar production is estimated at approximately 3,000 tons and carrageenan production at nearly 30,000 tons. To this latter figure should be added semi-refined carrageenan production, which is estimated at nearly 18,000 tons (Perez 1997).

Algae are a marine resource with various food applications. They can be incorporated into our diet directly as sea vegetables, intermediate food products or ingredients. Their consumption in this form has been traditional in Asia for thousands of years. Paradoxically, algae have been part of the Western diet for decades in the form of food additives (phycocolloids). This indirect introduction into our diet may lead us to reconsider Western consumers' perception of algae.

Algae family	Genus
Brown algae (alginates)	*Macrocystis* *Laminaria* *Ascophyllum* *Ecklonia* *Eisenia* *Nereocystis* *Sargassum*
Red algae (carrageenans)	*Chondrus* *Gigartina* *Eucheuma* *Hypnea* *Iridaea* *Gymnogongrus* *Ahnfeltia* *Furcellaria* *Kappaphycus*
Red algae (agar)	*Gelidium* *Gracilaria* *Pterocladia*

Table 6.19. *Algae used for the production of phycocolloids (Glicksman 1987)*

6.8. References

Aakre, I., Solli, D.D., Markhus, M.W., Mæhre, H.K., Dahl, L., Henjum, S., Alexander, J., Korneliussen, P.A., Madsen, L., Kjellevold, M. (2021). Commercially available kelp and seaweed products – Valuable iodine source or risk of excess intake? *Food & Nutrition Research*, 65, 10-29219.

Arasaki, S., Arasaki, T. (1985). *Les légumes de mer. Comment être et paraître en forme Japan Publications Trading*. Editions de la Maisnie, Douarnenez.

Cardoso, C., Valentim, J., Gomes, R., Matos, J., Rego, A., Coelho, I., Delgado, I., Motta, C., Castanheira, I., Prates, J.A.M. et al. (2024). Mackerel and seaweed burger as a functional product for brain and cognitive aging prevention. *Foods*, 13(9), 1332. doi: 10.3390/foods13091332.

Conte, É., and Payri, C. (2002). La consommation des algues en Polynésie française : premiers résultats d'une enquête. *Journal de la Société des Océanistes*, (114–115), 165–172.

da Costa, E., Melo, T., Reis, M., Domingues, P., Calado, R., Abreu, M.H., Domingues, M.R. (2021). Polar lipids composition, antioxidant and anti-inflammatory activities of the Atlantic red seaweed. *Grateloupia turuturu*. *Marine Drugs*, 19(8), 414.

Delaney, A., Frangoudes, K., Ii, S.A. (2016). Society and seaweed: Understanding the past and present. In *Seaweed in Health and Disease Prevention*, Fleurence, J. and Levine, I. (eds). Academic Press. Elsevier, Amsterdam.

Domínguez-González, M.R., Chiocchetti, G.M., Herbello-Hermelo, P., Velez, D., Devesa, V., Bermejo-Barrera, P. (2017). Evaluation of iodine bioavailability in seaweed using in vitro methods. *Journal of Agricultural and Food Chemistry*, 65(38), 8435–8442.

Dopter, A., El Ouadrhiri, Y., Margaritis, I. (2021). La consommation d'algues peut-elle conduire à un excès d'apport en iode ? *Cahiers de Nutrition et de Diététique*, 56(4), 220–225.

Fernandes, W.D., Pinto, F.R., Barroso, S., Gil, M.M. (2023). Development, characterisation, and consumer acceptance of an innovative vegan burger with seaweed. *Sustainability*, 15(14), 10869.

Fleurence, J. (1991). L'habilitation des algues en alimentation humaine. Le point sur la réglementation française. *Industries alimentaires et agricoles*, 6, 501–502.

Fleurence, J (2016). Seaweed as food. In *Seaweed in Health and Disease Prevention*, Fleurence, J. and Levine, I. (eds). Academic Press Elsevier, Amsterdam.

Fleurence, J. (2018). *Les algues alimentaires : bilan et perspectives*. Lavoisier Tec & Doc, Paris.

Glicksman, M. (1987). Utilization of seaweed hydrocolloids in the food industry. *Hydrobiologia*, 151(152), 31–47.

Marfaing, H. (2017). Qualités nutritionnelles des algues, leur présent et futur sur la scène alimentaire. *Cahiers de Nutrition et de Diététique*, 52(5), 257–268.

Mouritsen, O.G., Dawczynski, C., Duelund, L., Jahreis, G., Vetter, W., Schröder, M. (2013). On the human consumption of the red seaweed Dulse (*Palmaria palmata* (L.) Weber & Mohr). *Journal of Applied Phycology*, 25, 1777–1791.

Nang, H.Q. and Dinh, N.H. (1998). The seaweed resources of Vietnam. In *Seaweed Resources of the world*, Critchley, T. and Ohno, M. (eds). Japan International Cooperation Agency, Tokyo.

Nisizawa, K., Noda, H., Kikuchi, R., Watanabe, T. (1987). The main seaweed foods in Japan. *Hydrobiologia*, 151(152), 5–29.

Ohno, M. and Largo, D.B. (1998). The seaweed resources of Japan. In *Seaweed Resources of the World*, Critchley, T. and Ohno, M. (eds). Japan International Cooperation Agency, Tokyo.

Perez, R. (1997). *Ces algues qui nous entourent*. Ifremer, Plouzané.

Prabhasankar, P., Ganesan, P., Bhaskar, N., Hirose, A., Stephen, N., Gowda, L.R., Hosokawa, M., Miyashita, K. (2009). Edible Japanese seaweed, Wakame, (*Undaria pinnatifida*) as ingredient in pasta: Chemical, functional and structural evaluation. *Food Chemistry*, 115, 501–508.

Sauvageau, C. (1920). *Utilisation des algues marines*. Editions Doin G., Paris.

Sohn, C.H. (1998). The seaweed resources of Korea. In *Seaweed Resources of the World*, Critchley, T. and Ohno, M. (eds). Japan International Cooperation Agency, Tokyo.

Yamada, Y. (1941). Notes on some Japanese algae IX. Scientific Papers of the Institute of Algological Research, Faculty of Science, Hokkaido Imperial University. 2, 195–215.

Yeh, T.S., Hung, N.H., Lin, T.C. (2014). Analysis of iodine content in seaweed by GC-ECD and estimation of iodine intake. *Journal of Food and Drug Analysis*, 22(2), 189–196.

Zava, T.T. and Zava, D.T. (2011). Assessment of Japanese iodine intake based on seaweed consumption in Japan: A literature-based analysis. *Thyroid Research*, 4(1), 14.

Zemke-White, W.L. and Ohno, M. (1999). World seaweed utilisation: An end-of-century summary. *Journal of Applied Phycology*, 11, 369–373.

7

Regulations Applicable to Algae and Their Use in Food

7.1. Background

The harvesting and use of algae can be subject to specific regulations. This is particularly the case in France. The harvesting of marine plants has long been governed by Decree 90-719, which set the conditions for fishing, harvesting or collection of this resource (Arzel 1998).

In this decree, seaweed or kelp is classified into three distinct categories:

– shore seaweed;

– seaweed growing in the sea;

– stranded seaweed.

However, this decree was repealed on January 1 2015[1].

The main regulatory provisions concerning seaweed farming are now incorporated into the Rural and Maritime Fishing Code[2] and into the prefectural decrees governing the harvesting of shore seaweed in Brittany.

1. See: https://www.legifrance.gouv.fr/loda/id/JORFTEXT000000166991/2021-04-28/.
2. See: https://www.legifrance.gouv.fr/codes/section_lc/LEGITEXT000006071367/LEGISCTA.

Marine Algae, A Sustainable Food Resource,
by Joël FLEURENCE.

7.2. Regulations on seaweed farming

7.2.1. *French regulations on the harvesting of shore seaweed*

Kelp is an alga that remains attached to the substrate (rocks, sandy bottom) and is harvested on foot on the foreshore during low tide. It is distinct from algae washed up on beaches or from shipwrecks, which has been torn from its substrate by the force of currents, waves and storms. It is also very different from algae growing in the sea, which cannot be reached on foot.

The regulations governing the harvesting of shore seaweed are often based on local customs. These regulations may be regional or even departmental in nature[3]. From a regulatory point of view, shore seaweed harvesting is often considered a recreational activity. This is important because it limits this type of harvesting to personal use by the harvester. The harvested seaweed cannot be sold. Shore seaweed harvesting is permitted all year round, with the exception of certain species (Table 7.1). However, it is subject to time restrictions. It must be carried out between sunrise and sunset.

In addition to the aspects mentioned above, other criteria may be taken into account by the regulations. Among these, it should be noted that the harvesting of algae is prohibited except in the case of Laminaria. In other cases, the regulations stipulate a cutting height above the holdfast that varies according to species (Table 7.2).

Algae species	Authorized period
Chondrus crispus (Pioca)	May 1 to October 31
Mastocarpus stellatus	May 1 to October 31
Porphyra spp. (Nori)	May 1 to November 15
Palmaria palmata (Dulse)	May 1 to December 31

Table 7.1. *Authorized periods for harvesting red algae frequently found on the Brittany coast*

3. See: https://www.finistere.gouv.fr/contenu/telechargement/45341/321594/file/Algues%20de%20rive.pdf.

Algae species	Cutting height
Porphyra spp. (Nori)	25 cm
Palmaria palmata (Dulse)	25 cm
Himanthalia elongata (sea spaghetti)	80 cm
Saccharina latissima (*Laminaria saccharina*)	150 cm
Ascophyllum nodosum	30 cm

Table 7.2. *Authorized cutting height from the holdfast or base of the stipe*

7.2.2. *French regulations on the harvesting of algae growing in the sea*

Kelp growing in the sea refers to algae attached to seabeds that are not accessible on foot at low tide during equinox tides or spring tides. In France, the main species concerned are *Laminaria digitata* and *Laminaria hyperborea*. Algae drifting with the waves are also included in this definition.

This resource may only be harvested by vessels holding a fishing or marine farming license. This activity may only be carried out on the mainland coast between May 1 and October 30. Occasional restrictions may also apply to this type of fishing, such as:

– permanent or temporary bans on fishing in certain areas;

– limitations on the tonnage of certain species that may be caught;

– quantity limits per seaweed harvesting vessel;

– restrictions or even bans on the use of certain fishing gear.

If the resource is harvested by swimming or diving, the captain of the vessel must have signed a declaration of professional underwater fishing.

7.2.3. *French regulations on the harvesting of stranded algae*

Regulations on this type of seaweed harvesting are governed by two very brief articles of the French Rural Code (R922-42 and R922-43). The first prohibits the establishment of algae fisheries by means of stakes or any other method. The second states that for policing reasons and in consultation with the mayors concerned, the administrative authority may decide to organize the collection of algae. The algae concerned are mainly invasive species, such as Ulva in Brittany and Sargassum in

Normandy and the West Indies. This article is therefore part of the fight against green tides (*Ulva armoricana*) and brown tides (*Sargassum* spp.).

7.2.4. *Specific provisions for the Mediterranean*

The harvesting of algae growing in the salt ponds of the Mediterranean is subject to the regulatory provisions mentioned above.

7.2.5. *Foreign regulations*

France is the leading algae producer in Europe. Its regulations on seaweed farming are often more precise and detailed than those of other countries. Many of these foreign regulations incorporate and formalize traditional practices.

In Ireland and the United Kingdom, the foreshore may be privately owned and, unlike in France, does not always belong to the public domain. The rights to use this part of the coastline are often the prerogative of landowners. These rights include the possibility of harvesting algae in the area concerned.

In 1933, Ireland introduced highly restrictive regulations known as the Foreshore Act. These regulations stipulate that any person or company wishing to harvest must obtain a permit from the Department of Environment, Heritage and Local Government (Frangoudes et al. 2012)[4].

Since 1933, however, these regulations have evolved and certain aspects are now better defined, such as the regulatory definition of the foreshore and its ownership:

– "The foreshore of Ireland is classed as the land and seabed between the high water of ordinary or medium tides [...] and the twelve-mile limit (12 nautical miles equals approximately 22.24 kilometers)";

– "Foreshore also covers tidal areas of rivers particularly estuaries".

The most interesting regulatory change concerns the ownership of the foreshore:

> All the foreshore of Ireland is presumed to be owned by the State unless valid alternative title is provided.

The new adjustments to the law therefore still recognize the potentially private nature of the foreshore and the need, as in 1933, to obtain a license authorizing seaweed farming. In the United Kingdom, seaweed farming is also subject to private

4. See: https://www.gov.ie/en/publication/4ddcf-foreshore-your-questions-answered/.

law. This is because the foreshore is often privately owned. Private owners must grant farmers access to it.

In Scotland, this right is subject to a lease between the farmers and the owners. In addition, farming seaweed beyond the low tide line requires an additional permit. This permit must be obtained from the Crown Estate. The number of permits issued is very limited, and the area of application may in some cases be confidential (Frangoudes et al. 2012).

In Norway, seaweed farming, particularly *Ascophyllum nodosum* (Kelp), takes place in areas of the foreshore considered private property. In this case, as previously mentioned, farmers must obtain permission from the landowners. However, *Laminaria hyperborea,* a brown alga used for the production of alginates, has public regulatory status. This species grows in deep waters and is therefore harvested in the public maritime domain. Its collection therefore requires authorization from the Norwegian government.

French regulations on seaweed farming differ from those of other European countries in terms of the foreshore's legal status. In France, the foreshore belongs to the public domain and its use is the responsibility of public authorities (prefectures, government).

In many countries, the foreshore may be considered private property. In this case, its use is governed by agreements or leases between harvesters and owners. When the latter falls under public domain, depending on the circumstances (Ireland, Norway), seaweed farming requires prior authorization.

More generally, seaweed farming practices vary from country to country and even from region to region. They may or may not be subject to specific regulations.

7.3. Regulations on the use of algae in food

The use of algae in human food may be subject to no regulation, general food regulations or very specific regulations. The latter is the case in France.

7.3.1. *French regulations*

Until the end of the 1980s, algae consumption in France was extremely marginal and confined to a very small number of enthusiasts, who associated it with periods of famine.

The restructuring of the algae industry in Brittany, which was mainly dedicated to the production of alginates, was based on the emergence of a new industry, namely that of "edible algae". An economic interest group, GRAAL (Regional Group for Edible Algae), was responsible for promoting and developing this new industry.

Although algae have been consumed by humans for thousands of years, the French government (Directorate General for Consumer Affairs and Fraud Control) classified them as a non-traditional food. As such, the marketing of algae as an occasional vegetable or ingredient could only be considered after consulting the French High Council for Public Health (CSHPF) and the National Academy of Medicine (Fleurence 1991).

In 1990, the Bulletin of the Ministry of Social Affairs (text 1705, p. 103) transcribed the favorable opinion of these bodies for the marketing of 11 species of macroalgae and Spirulina as occasional vegetables or ingredients. Laminaria, the first macroalgae to receive a favorable opinion in 1984, was excluded from the list of algae authorized for human consumption in 1990 due to its high iodine content (5,000–10,000 mg/kg of DM). This regulatory reversal was fatal at the time for the development of products containing Laminaria such as intermediate food products (Chapter 6, section 6.3). Currently, the list of algae authorized for human consumption includes 24 species, one of which, *Spirulina* sp., is not an alga but a cyanobacterium (Table 7.3).

In conjunction with this list of algae authorized for human consumption, a number of mineral and microbiological toxicity criteria have also been established. In France, the maximum heavy metal and iodine content of edible algae was set by the CSHPF in 1988. In 2007, the AFSSA (French Food Safety Agency) ruled on the maximum levels of mineral (inorganic) arsenic and iodine in laminaria algae intended for human consumption. The mineral toxicity criteria currently applicable to all edible algae are listed in Table 7.4.

	Scientific name of the species	Common name
Brown algae	*Ascophyllum nodosum*	Kelp
	Fucus serratus, Fucus vesiculosus	Wrack or black kelp
	Himanthalia elongata	Spaghetti or sea bean
	Laminaria digitata	Breton Kombu
	Laminaria saccharina	Royal Kombu
	Laminaria japonica	Kombu
	Undaria pinnatifida	Wakame
	Alaria esculenta	Atlantic Wakame

Red algae	*Palmaria palmata*	Dulse
	Porphyra umbilicalis	Nori
	Porphyra tenera	Nori
	Porphyra yezoensis	Nori
	Porphyra dioica	Nori
	Porphyra purpurea	Nori
	Porphyra laciniata	Nori
	Porphyra leucostica	Nori
	Chondrus crispus	Pioca, Lichen
	Gracilaria verrucosa	Ogonori
	Lithothamnium calcareum	Maerl
Green algae	*Ulva* sp.	Sea lettuce
	Enteromorpha sp.	Aonori
Microalgae	*Odontella aurita*	–
	Chlorella sp.	Chlorella
	Spirulina sp. (Cyanobacteria)	Spirulina

Table 7.3. *Updated list of algae (macro- and microalgae) and cyanobacteria authorized in France for consumption as sea vegetables or ingredients (Fleurence 2016)*

Minerals	Maximum content (mg.kg^{-1} of dry matter)
Mineral arsenic (inorganic)	3
Cadmium	0.5
Lead	5
Tin	5
Mercury	0.1
Iodine	2,000

Table 7.4. *Maximum levels of toxic minerals and iodine in algae authorized as occasional vegetables or ingredients in France (Fleurence 2016)*

Arsenic is an element taken into account by French regulations, but in a different way from other countries (Box 7.1). Only mineral or inorganic arsenic is covered by the regulations (Table 7.4).

French regulations take into account mineral arsenic and not total arsenic, as is the case with the FDA (Food and Drug Administration) for brown algae marketed in the United States under the name "Kelp". This difference in assessment is the result of a different perception of the risk associated with the presence of arsenic between American and French experts.

Furthermore, mineral arsenic is a misnomer. It is actually inorganic arsenic, in other words free arsenic, since it is not bound to an organic molecule. In many algae products, arsenic can be found in combination with organic compounds such as sugars or betaine. These compounds are known as arsenobetaines or arsenosugars. Studies on the in vivo toxicity of arsenic present in Nori and "Kelp" have shown that organic arsenic was mainly in the form of arsenosugars, which were mainly excreted in the urine between 10 and 60 hours after ingestion of Nori, and between 17 and 37 hours after consumption of "Kelp" (Le et al. 1994).

In view of the available research, French experts therefore considered that only free arsenic, not complexed with an organic molecule, was potentially toxic and should be included in the definition of mineral toxicity criteria. This seemed reasonable, as the differences between total arsenic and inorganic arsenic levels in brown algae appear to be very high (Edmonds and Francesconi 1993) (Table 7.5).

Box 7.1. *The special case of inorganic arsenic, a unique feature of French regulations*

Algae and seafood	Total arsenic (mg.kg^{-1})	Organic arsenic (mg.kg^{-1})	Inorganic arsenic (mg.kg^{-1})	% inorganic arsenic
Undaria pinnatifida	1.6	1.2	0.14	8.7
Hizikia fusiformis	9.3	1.8	5.00	54
Laminaria hyperborea	14.2	13.9	0.10	0.7
Laminaria japonica	2.5	2.00	0.08	3.2
Ecklonia cava	6.1	3.0	0.6	9.8
Ecklonia radiata	9		0.36	4.0
Sargassum bracteolosum	7.9	–	0.17	2.2
Pecten alba (1)	7.2	–	0.04	0.6
Pandalus borealis (2)	6.0	5.9	0.5	8.3
Mytilus edulis (3)	2.8	2.5	0.3	10.7
Gadus morhua (4)	–	4.6	0.3	6.1
Scomber scombrus (5)	0.9	0.8	0.4	44
Penaeus plebejus (6)	–	5.8	0.08	1.4
Jasus novaehollandiae (7)	12.4	–	0.09	0.7

(1) Scallops. (2) Northern prawn. (3) Mussels. (4) Cod. (5) Mackerel. (6) Australian king prawn. (7) Australian rock lobster.

Table 7.5. *Total inorganic and organic arsenic levels in selected edible brown algae and other marine products (Edmonds and Francesconi 1993)*

However, it is not easy to apply regulations to inorganic arsenic in algae, as there is no standard method for measuring this element in the algal matrix (Box 7.2). In addition to mineral toxicity criteria, French regulations also take into account a number of microbiological criteria (Box 7.2 and Table 7.6).

Inorganic arsenic in algae is currently measured after acid extraction (HCl) and hydride generation by atomic absorption spectrophotometry, according to a standard method (NF-EN-15517). This is worth noting because, for a long time, the lack of a standardized analytical method for assessing the inorganic arsenic content led to a conflict over the validity of this particular mineral toxicity criterion.

In addition to criteria related to mineral toxicity, microbiological criteria are also applied to dried algae packaged in bags (Table 7.6). These criteria are similar to those required for vegetables and do not include testing for potentially pathogenic bacteria of marine origin such as *Vibrio*. Certain products, such as lactofermented, pasteurized or frozen algae, are not subject to the same microbiological criteria as dried algae (Table 7.6). Only the microbiological criteria traditionally applied to foodstuffs appear to be applicable to these particular algae-based products.

Box 7.2. *Practical limits of application of the regulatory context*

Microbiological contaminants	Levels
Mesophilic aerobic flora	≤ 100,000 germs/g
Fecal coliforms	≤ 10 germs/g
Anaerobic sulfite-reducing flora	≤ 100 germs/g
Staphylococcus aureus	≤ 100 germs/g
Clostridium perfringens	≤ 1 germ/g
Salmonella	Absent

Table 7.6. *Microbiological criteria applied to dried algae used in human food in France (Fleurence 2016)*

French regulations are based on European regulations in a specific area of algae use in food, namely food supplements[5]. The maximum levels of toxic minerals differ from those recommended for the consumption of algae as a vegetable or condiment (Table 7.7). This is particularly true for cadmium, where the maximum acceptable level is 3 mg.kg^{-1} of marketed product for algae-based food supplements. By way of

5. See: https://eur-lex.europa.eu/legal-content/EN/TXT/?uri=CELEX:32021R1323.

comparison, the maximum permitted level for algae vegetables is 0.5 $mg.kg^{-1}$ of dry matter. This expression of levels according to different reference criteria, namely raw product or dry product, does not allow a qualitative comparison of cadmium levels between the two regulations. Nevertheless, it should be noted that the maximum cadmium content permitted for algae-based food supplements is significantly higher than that recommended for food supplements not containing algae (Table 7.8).

Element	Maximum content ($mg.kg^{-1}$ of marketed product)
Cadmium	3
Mercury	0.1
Lead	3

Table 7.7. *Maximum levels of toxic minerals in food supplements containing dried seaweed*

Product	Maximum cadmium content ($mg.kg^{-1}$)
Dietary supplement	1.0
Food supplement containing algae	3.0

Table 7.8. *Comparison of recommended maximum cadmium for dietary supplements containing or not containing algae in their composition*

More generally, a decree known as the "plant decree"[6] includes certain species of algae already recognized in the food list for the production of food supplements.

It also adds certain species not included in the list of algae authorized for human consumption (Table 7.9) for this type of application.

French regulations on the use of algae in direct human consumption as a vegetable, ingredient or condiment are currently among the most detailed and restrictive in Europe. However, they have the advantage of being specific to algae and of establishing health, safety and quality criteria that are selling points for the

6. See: https://www.legifrance.gouv.fr/loda/id/LEGITEXT000029255041/.

export of French products to more knowledgeable and demanding markets for this type of foodstuff.

Type of algae	Species	Commercial name
Macroalgae	*Gelidium corneum*	Agar-agar
Macroalgae	*Laminaria hyperborea*	Hyperborean kelp
Macroalgae	*Macrocystis pyrifera*	Kelp
Macroalgae	*Mastocarpus stellatus*	–
Macroalgae	*Sargassum fusiforme*	Hai zao, Hiziji
Microalgae	*Aphanizomenon* sp.	Blue-green algae from Klamath Lake
Microalgae	*Dunaliella salina*	–
Microalgae	*Haematococcus pluvialis*	–

Table 7.9. *List of algae (macroalgae and microalgae) included in the "Plants Decree" that are not included in the list of algae authorized as vegetables but which may be used as food supplements*

7.3.2. *European regulations*

Unlike France, there are no European regulations specifically applicable to edible algae. Algae were initially included in European Regulation (EC) No. 258/97 published in 1997, known as the Novel Food Regulation. This regulation on novel foods and food ingredients mentions algae in paragraph 2 of Article 1 (EC 1997):

> This regulation shall apply to the placing on the market within the Community of foods and food ingredients which have not hitherto been used for human consumption to a significant degree within the Community and which fall under the following categories:
>
> [...] d) foods and food ingredients consisting of or isolated from micro-organisms, fungi or algae.

This regulatory provision dating from 1997 has been repealed and replaced by Regulation (EU) 2015/2283. As previously, algae-based food products are classified as novel foods by the 2015 European Directive[7]. By classifying algae as novel

7. See: https://eur-lex.europa.eu/legal-content/EN/TXT/?uri=CELEX%3A32015R2283.

foods, European regulations implicitly recognize the non-traditional nature of this food resource. In doing so, they validate the regulatory choice initially made by France.

However, European regulations do not establish a positive list of algae species authorized for human consumption. This last point distinguishes them from French regulations, which therefore appear more restrictive. The text of the European regulation also does not refer to toxicological criteria applicable to edible algae and thus differs from French regulations. The European regulation includes algae in a group of foods that are of marginal consumption in Europe or are of more recent conception (foods derived from genetically modified organisms). It therefore appears less specific and less restrictive for the use of algae in food than the French regulation.

7.3.3. *US regulations (United States)*

The health criteria applicable to edible algae are defined in the United States by the Food Chemical Codex. In 1972, these criteria applied to "Pacific kelp", namely the brown alga *Macrocystis pyrifera* and related species. This text defined mineral toxicity criteria for arsenic, lead and iodine (Table 7.10). It also provided guidance on certain product characteristics to be met, such as ash content and dry matter content (Table 7.10).

Criterion	Food Chemical Codex *1972*	Food Chemical Codex *2003*
Dry matter	≤ 13% FM	≤ 13% DM
Ash	≤ 35% DM	≤ 45% DM
Iodine	0.15-0.22% DM	0.1-0.5%
Arsenic	≤ 3 mg.kg^{-1} of DM	≤ 1 mg.kg^{-1} of DM
Lead	≤ 10 mg.kg^{-1} of DM	≤ 2 mg.kg^{-1} of DM
Heavy metals	≤ 40 mg.kg^{-1} of DM	–

DM: dry matter; FM: fresh matter

Table 7.10. *Recommended quality criteria for* Macrocystis pyrifera *(Kelp) by the Food Chemical Codex in 1972 and 2003*

The text also provided information on storage conditions (protected from light) and the intended use of the algae concerned (food supplement). In 2003, the Food Chemical Codex amended some of the mineral toxicity criteria applicable to algae

(Table 7.10) and specified the species concerned by this type of application. The 2003 text extends the concept of edible algae to all brown algae species related to *Macrocystis* and all species of the genus *Laminaria*.

The 2003 Food Chemical Codex recommendations applicable to edible algae are stricter than those initially established in 1972 for arsenic and lead. They are also stricter for these two elements than those currently in force in French regulations (Table 7.4). By way of comparison, it is interesting to note that the maximum arsenic content of 3 $mg.kg^{-1}$ of dry algal matter included in French regulations originates from the value recommended by the Food Chemical Codex in 1972 (personal communication, Joël Fleurence).

Apart from the Food Chemical Codex, which defines the health quality criteria that edible algae must meet, algae are subject to specific regulations (Code of Federal Regulations 2017a, 2017b). These regulations establish a list of algae recognized as safe for human consumption. This list complements and clarifies the one initially defined in the Food Chemical Codex. This approach is similar to that of French regulations, which establish a list of algae species that can be consumed.

The US regulations mainly define the use of algae as an ingredient or intermediate food product, whether in dried, powdered or flake form. Two large families are taken into account from a regulatory standpoint: brown algae (Table 7.11) and red algae. The species covered by these regulations are listed, and the geographical areas where they may be harvested are also specified in the regulatory text.

Thus, brown algae authorized by US regulations must be harvested in the coastal waters of the North Atlantic or Pacific Ocean. This geographical area is large enough to include species native to the northern European and Japanese coasts in the list of algae concerned. The food applications covered by the regulations are flavorings, seasonings and spices (Table 7.11). Algae are therefore considered food ingredients under US regulations.

For brown algae, as for red algae, the regulations do not set a maximum level for incorporation into the finished product. It is simply stipulated that the level of algae incorporation must not exceed that of a conventional food ingredient, as defined by good manufacturing practices.

Brown algae	Spices	Flavorings	Seasonings
Analipus japonicus	+	+	+
Eisenia bicyclis	+	+	+
Hizikia fusiformis	+	+	+
Kjellmaniella gyrata	+	+	+
Laminaria angustata	+	+	+
Laminaria claustonia	+	+	+
Laminaria digitata	+	+	+
Laminaria japonica	+	+	+
Laminaria longicruris	+	+	+
Laminaria longissima	+	+	+
Laminaria ochotensis	+	+	+
Laminaria saccharina	+	+	+
Macrocystis pyrifera	+	+	+
Petalonia fascia	+	+	+
Scytosiphon lomentaria	+	+	+
Undaria pinnatifida	+	+	+

Table 7.11. *List of brown algae authorized for human consumption and food applications (Code of Federal Regulations 2017a, section 184.1120)*

There are seven species of red algae authorized by US regulations (Code of Federal Regulations 2017b). These are *Gloiopeltis furcata*, *Porphyra crispata*, *Porphyra dentata*, *Porphyra perforata*, *Porphyra suborbiculata*, *Porphyra tenera* and *Rhodymenia palmata* (*Palmaria palmata*). These algae must be harvested from the coasts of the South Pacific, Japan, China, Korea, Taiwan or the east and west coasts of the United States, depending on the geographical location of the species. The dry matter content of these algae must not exceed 20% and the maximum iodine content must not exceed 500 $mg.kg^{-1}$ of dry matter.

7.3.4. *Other regulations*

In Australia and New Zealand, there are recommendations on algae consumption for pregnant and breastfeeding women, and children. There are no binding regulations, but simply a reminder about the excessive iodine intake that could result from excessive consumption of brown algae. In this context, it is recommended that children do not consume algae of the genera *Laminaria*, *Undaria*, *Sargassum*, *Alaria*, *Eisenia*, *Fucus* or *Ascophyllum* more than once a week (Food Standards Australia New Zealand 2023).

Iodine remains the main concern for Australian authorities. It shared the results of a study published in 2024 on the Food Standards Australia New Zealand website[8]. According to this study, most algae are considered safe to eat despite their iodine content. This is particularly true of Wakame (*Undaria pinnatifida*) and Nori (*Porphyra* spp.). However, algae with a high iodine content are classified as unfit for human consumption. This includes imported brown algae with an iodine concentration of more than 1,000 mg of iodine per kg of dry matter.

Inorganic arsenic is also a criterion taken into account by the Australian authorities. The maximum recommended level is 1 $mg.kg^{-1}$ of dry matter. This level is three times lower than that imposed by French regulations (Table 7.4). A study of algae products sold in Australia shows that Hijiki (*Sargassum fusiforme*) may exceed this level and its frequent consumption is therefore considered a risk by the authorities.

The Australian regulatory approach is mainly based on recommendations and risk assessments related to the consumption of certain algae. Unlike French regulations, it is not based on formal authorizations or prohibitions. Similarly, it does not establish a positive list of algae species that can be consumed as sea vegetables or condiments. However, it does publish a list of brown algae that can be consumed in various culinary preparations (soups, salads, etc.) (Table 7.12). These algae[9] are classified as safe by the Australian authorities, provided that they comply with the maximum levels of iodine and inorganic arsenic mentioned above.

Countries where algae are traditionally consumed (China, Korea, Japan) have no specific regulations for the use of this marine resource in human food. Commercially available products must simply meet the regulatory requirements applicable to foodstuffs. Although Japan does not apply mineral toxicity criteria to its own edible algae, it imposes extremely strict heavy metal limits on imported marine plants (Perez 1997).

---

8. See: https://www.foodstandards.gov.au/sites/default/files/science-data/surveillance/Documents/Iodine%20in%20Seaweed.pdf.
9. See: https://www.foodstandards.gov.au/consumer/safety/brownseaweed/brownseaweedstable.

Algae species	Common name	Uses
Laminaria spp. *Laminaria japonica*	Kelp, Kombu, Konbu, Suboshi-Kombu, Aokombu, Haidai, Dashima, Oarweed, Wild kelp, Fingered tangle, Sea tangle	Dashi (broths), soups, salads, drinks, sauces, sashimi, kombucha (seaweed tea)
Undaria spp.	Wakame, Haiboshi-Wakame, Quandai-cai, Miyeok, Sea fern, Mekab	Miso soups, salads
Sargassum fusiforme, *Sargassum* spp.	Hiziki, Hoshi-Hiziki	Sea vegetables, soups, stir-fries
Cladosiphon okamuranus	Mozuku	Salads, ingredients
Alaria esculenta	Winged kelp, Dabberlocks, Badderlocks, Atlantic Wakame	Salads, sea vegetables
Eisenia bicyclis	Arame, Kelp	Sea vegetables, salads, garnishes, condiments
Ecklonia maxima	Sea bamboo, Paddle grass	–
Fucus vesiculosus	Fucus, cut grass	Additives, flavorings, dietary supplements
Durvillaea antarctica	Bull kelp, Cochayuyo, Hulpe	Stews, salads
Ascophyllum nodosum	Norwegian kelp, Knotted wrack, Egg wrack	Alginates, sea vegetables
Postelsia palmaeformis	Sea palm	Chinese dishes

Table 7.12. *List of brown algae and associated preparations available in Australia and considered safe for consumption*

7.4. References

Arzel, P. (1998). *Les laminaires sur les côtes bretonnes. Évolution de l'exploitation et de la flottille de pêche, état actuel et perspectives*. Ifremer, Plouzané.

Code of Federal Regulations (2017a). Direct food substances affirmed as generally recognized as safe. Brown algae. Title 21, volume 31, section 184.1120.

Code of Federal Regulations (2017b). Direct food substances affirmed as generally recognized as safe. Red algae. Title 21, volume 31, section 184.1121.

Edmonds, J.S. and Francesconi, K.A. (1993). Arsenic in seafoods: Human health aspects and regulations. *Marine Pollution Bulletin*, 12, 665–674.

European Commission (1997). Regulation (EC) No 258/97 of the European Parliament and of the Council of 27 January 1997 concerning novel foods and novel food ingredients [Online]. Available at: https://eur-lex.europa.eu/eli/reg/1997/258/oj/eng.

Fleurence, J. (1991). L'habilitation des algues en alimentation humaine : le point sur la réglementation française. *Industries Alimentaires et Agricoles*, 6, 501–502.

Fleurence, J. (2016). Seaweed as food. In *Seaweed in Health and Disease Prevention*, Fleurence, J. and Levine, I. (eds). Academic Press, Elsevier, Amsterdam.

Food Standards Australia New Zealand (2023). Pregnancy and healthy eating [Online] https://www.foodstandards.gov.au/consumer/nutrition/pregnancy/healthy-eating.

Frangoudes, K., Jacob, C., Lesueur, M., Mesnildrey, L., Spinec, F., Folliard, G., Pien, S., Rebours, C., Meland, M., Gallastegi, I. et al. (2012). Guide pour une meilleure gestion de la filière macro-algues en Europe [Online]. Available at: http://www.netalgae.eu/uploadedfiles/NE_guide_fr.pdf.

Le, X.C., Cullen, W.R., Reimer, K.J. (1994). Human urinary arsenic excretion after one-time ingestion of seaweed, crab, and shrimp. *Clinical Chemistry*, 40(4), 617–624.

Perez, R. (1997). *Ces algues qui nous entourent*. Ifremer, Plouzané.

8
Outlook

8.1. Background

Marine algae, whether microalgae or macroalgae, are primary producers of oxygen and are part of the first link in the oceanic food chain. Like terrestrial plants, they contribute to atmospheric carbon sequestration, and developing their cultivation at sea is likely to strengthen the fight against greenhouse gas emissions.

Furthermore, algae production, whether through cultivation or harvesting, does not require the use of fresh water, unlike water-intensive agricultural crops such as soybeans, corn and wheat. These crops are sources of protein used in human and animal nutrition. In the field of animal nutrition, they are used in the formulation of feed for livestock, as well as for fish farming.

However, certain species of algae, particularly red algae, can have protein contents equivalent to or even higher than those reported for soybeans (Fleurence and Guéant 1999).

In this context, the prospect of using algae as a complementary or even alternative source of protein, particularly for soy, in aquaculture is worth exploring. Similarly, developing the consumption of algae as a source of nutrients (proteins, minerals) in human food could help reduce the use of land-based animal products and thereby lower the ecological impact of livestock farming.

For a color version of all the figures in this chapter, see www.iste.co.uk/fleurence/marinealgae.zip

8.2. Algae: an alternative source of plant protein in human nutrition

Changing dietary habits with the emergence of vegetarian and vegan trends are reigniting the debate on the consumption of plants as an alternative source of protein for human nutrition.

Land plants are well suited to meeting this new demand for protein, particularly legumes such as beans, lentils and soybeans. Soybeans are characterized by a protein content of up to 36% of dry matter and contain all eight amino acids necessary for human nutrition. These amino acids, which are essential for adults, are tryptophan, lysine, methionine, leucine, isoleucine, threonine, phenylalanine and valine. However, soybeans are intensively cultivated in the United States and Brazil using genetically modified seeds, which poses a problem in meeting the demand of consumers who are increasingly aware about the origin of their food.

Algae currently consumed in traditional diets contain the eight essential amino acids mentioned above (Fleurence 2004; Fleurence et al. 2018). Some of them have significantly higher levels of sulfur amino acids, such as methionine, than those reported for ovalbumin, the reference dietary protein for the FAO (Chapter 4, section 4.3).

Red algae such as *Palmaria palmata* and *Porphyra tenera* have protein contents that can reach a maximum of 26–47% of dry algal matter (Chapter 4, section 4.3). The digestibility of these proteins currently appears to be the main obstacle to the use of algae as a protein source for human consumption, but processing methods such as enzymatic liquefaction or fermentation can significantly improve digestibility (Chapter 5, sections 5.7 and 5.8). Furthermore, protein digestibility varies depending on the species and season, and some red algae with a high protein content (*Gracilaria verrucosa*) do not require any prior treatment to improve their already satisfactory digestibility (Marrion et al. 2005).

As mentioned above, algae do not require fresh water for cultivation and appear to be an economical source of dietary protein from this point of view. By way of comparison, the production of one kilogram of soybeans requires 900 liters of fresh water (Table 8.1).

This aspect is important because it alone justifies the interest that should be shown in algae as a source of marine plant protein. The use of algae protein as a supplement, or even a substitute in some cases, for plant protein sources now appears to be an interesting prospect in the current ecological and societal context.

Plant	Volume of water required to produce 1 kg of plants
Soy	900 liters
Wheat	500 liters
Corn	454 liters
Seaweed	0 liters

Table 8.1. *Freshwater consumption of different crops (Fleurence 2018)*

8.3. Algae as functional foods: the specific case of hypoallergenic foods

The use of algae as a source of protein in human nutrition is not only hampered by issues relating to protein digestibility, assimilation and, ultimately, nutritional value. Food allergens are mainly protein-based, and seafood is often described as an allergenic food. Allergies to fish, crustaceans and shellfish are well known and often cited in the literature (Daul et al. 1993).

Paradoxically, algae are rarely mentioned in cases of food allergies resulting from the consumption of seafood. Only one serious case of food intolerance, followed by death, has been reported in Japan after consumption of Ogonori and tuna (Noguchi et al. 1994).

Algae, and more specifically macroalgae, have often been presented as having anti-inflammatory or anti-allergic properties (Fleurence and Ar Gall 2016). This characteristic is an asset for promoting algae in human nutrition and developing functional foods or food supplements that are free of major allergens. This highly theoretical concept of "hypoallergenic" foods therefore appears to be a promising avenue for research and the valorization of algae.

The anti-inflammatory and anti-allergic activities of algae are often associated with extracts rich in phenolic derivatives or polysaccharides. Among the phenolic derivatives found in algae are caffeic acid, better known as phenolic acid, certain quinones such as coumarins, trihydroxybenzenes such as phloroglucinol, and phlorotannins such as bromophloroglucinol (see Figure 8.1).

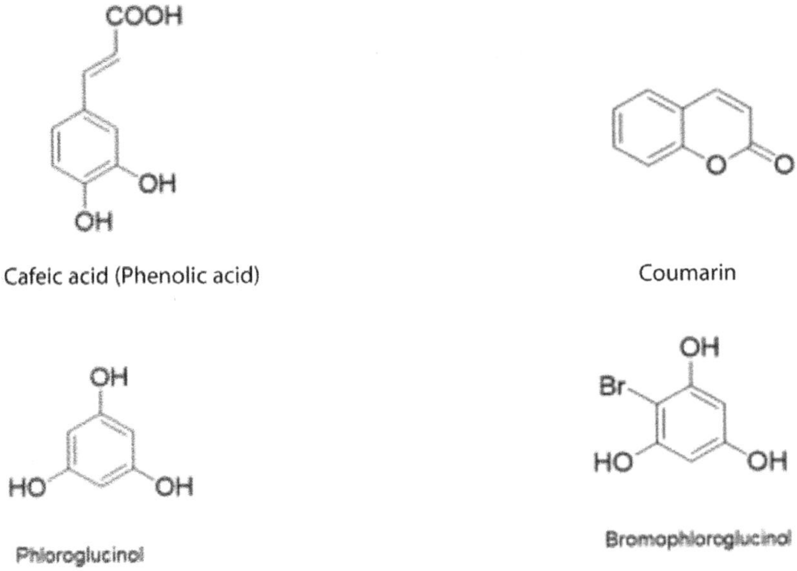

Figure 8.1. *Some phenolic derivatives present in algae extracts with anti-allergic activity (Fleurence and Ar Gall 2016; Fleurence 2025)*

The anti-allergic and anti-inflammatory mechanisms of algal extracts have mainly been studied using Phaeophyceae from the genera *Ecklonia*, *Eisenia*, *Sargassum*, *Laminaria* and *Undaria*.

Phenolic molecules isolated from *Ecklonia* and *Eiseina* algae inhibit hyaluronidase activity, an enzyme involved in inflammatory responses, particularly in the skin. Some of these modulating molecules isolated from *Ecklonia* have been named "eckols". These molecules belong to the phlorotannin family, along with fucols and fuhahols (see Figure 8.2) (Fleurence and Ar Gall 2016; Fleurence 2025).

Algal polysaccharides have also been described in the literature as having anti-inflammatory or anti-allergic properties. These are mainly sulfated polysaccharides such as porphyranes (Vo et al. 2012) and fucoidans (Ngo and Kim 2013). These polysaccharides are present in traditionally consumed algae such as Nori and Wakame.

The use of algae as a functional food is not limited to the development of products that do not contain major allergens. However, this avenue should be explored further, as there are cases of severe food allergies to animal proteins. In this context, certain species, due to their high protein content and complete amino acid profile, could become an alternative for the development of "health" foods. This type of functional food for people with severe food allergies must first be evaluated and approved by the relevant authorities.

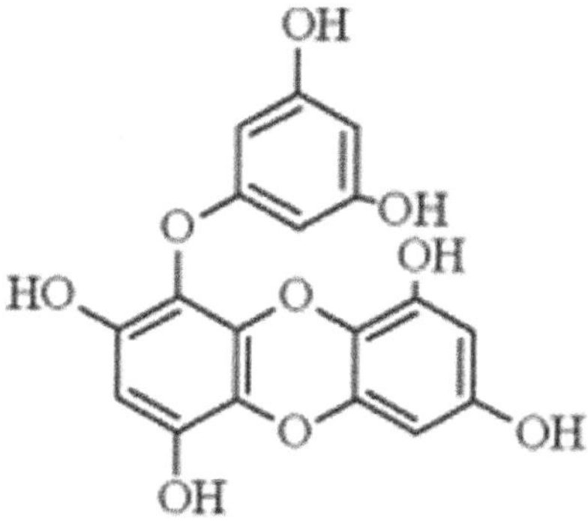

Eckol

Figure 8.2. *Eckol molecule present in brown algae extracts with anti-allergic activity (Fleurence and Ar Gall 2016)*

8.4. Other cases

Algae are often referred to as "health foods". Thus, the development of functional algae foods to maintain consumer health is a concept that is often discussed, whether in research (Wells et al. 2017), industry or even in the public domain.

It is undeniable that marine algae produce unique molecules that are not found in terrestrial plants and whose biological activities are of interest in the field of human (Fleurence 2025) and animal health. Among these molecules are mainly polysaccharides, which behave like dietary fibers (Chapter 4, section 4.2), but are also described in the literature as having antitumor, antithrombotic, hypocholesterolemic and antibacterial activities (Table 8.2).

Alginates derived from the algae *Saccharina latissima*, *Undaria pinnatifida*, *Laminaria digitata*, *Fucus vesiculosus* and *Ascophyllum nodosum* exhibit one or more of these activities (Holdt and Kraan 2011). Similarly, fucoidans from certain edible brown algae are often described for their anti-inflammatory, antioxidant, antitumor, anticoagulant and anti-aging properties (Holdt and Kraan 2011).

Polysaccharides	Biological activities	Species
Alginates	Antibacterial	*Saccharina latissima*
	Antitumor	*Undaria pinnatifida* *Sargassum vulgare*
Fucoidans	Anticoagulant	*Laminaria digitata*
	Anti-sclerotic	*Laminaria japonica*
	Anti-inflammatory	*Laminaria hyperborea*
	Antitumor	*Sargassum vulgare* *Sargassum horneri* *Undaria pinnatifida*
	Antioxidant	*Laminaria* sp.
	Prevention of Alzheimer's disease	*Laminaria* sp.
	Anti-aging	*Laminaria* sp.

Table 8.2. *Examples of biological activities of therapeutic interest associated with algal polysaccharides (Holdt and Kraan 2011)*

Fucoidans isolated from the brown algae *Ecklonia cava*, *Sargassum horneri* and *Costaria costata* have been described as having inhibitory activity on the proliferation of human malignant cells (melanoma, colon cancer) (Ermakova et al. 2011). Studies conducted on fucoidans obtained from the edible alga *U. pinnatifida* have also demonstrated the effectiveness of this type of molecule in inhibiting the viability of certain cancer cell lines (T-47D breast cancer, SK-MEL-28 melanoma). Fucoidans also appear to develop antitumor activity against prostate cancer with a 90% reduction in cell viability at a dose of 1,000 $\mu g.mL^{-1}$ (Gutiérrez-Rodriguez et al. 2017). Some of these fucoidans have been isolated and well characterized. This is the case with "Miyeokgui" (Miyeok, the Korean name for *U. pinnatifida*), a fucoidan which is a sulfated and acetylated galactofucan with an apparent molecular mass of 1,246 kDa. This fucan exhibits antitumor activity against prostate cancer, brain cancer and liver carcinoma (Synytsya et al. 2010).

In addition to polysaccharides, algae are also an interesting source of minerals that are beneficial to human health (iron, calcium, magnesium) (Chapter 4, section 4.5). In addition, certain species, such as the brown alga *U. pinnatifida,* have an interesting omega-3 polyunsaturated fatty acid profile (Chapter 4, section 4.4.2) and could contribute to cardiovascular protection in consumers in the same way as other foods (oily fish).

Despite all these nutritional benefits, the development of functional foods based on algae currently faces several challenges. The first is the variability in the content of nutrients of interest depending on the species, season and even geographical location (Wells et al. 2017). Secondly, the bioavailability of these nutrients, particularly minerals, is not always well understood. This does not facilitate the development of nutritional or "health" claims for products. The difference in regulatory status of algae used in human food in different countries also complicates the development of functional products, probably more so in Europe than in Asia or the United States.

However, there are some solutions that can be applied to overcome these difficulties. Firstly, the use of algae cultivated and harvested at favorable times in terms of their nutritional composition remains a realistic approach for certain species such as *U. pinnatifida*, *L. japonica* and *Porphyra tenera*. Secondly, studies should be carried out on the in vivo rather than in vitro digestibility of algae constituents and their bioavailability, as information in this area is still too piecemeal.

8.5. Algae: an alternative source of vegetable protein in animal feed

The introduction of algae into human food can also be considered indirectly, especially in Western countries where this foodstuff is not associated with traditional dietary practices. In light of this observation, it seems interesting to consider the possibility of using algae as a source of nutrients for animals raised for human consumption.

Protein is an essential component of animal nutrition, as it contributes significantly to the growth process. In Western countries, livestock farming relies mainly on soybeans in the form of seeds or cattle cake to provide the protein required for animal growth. Europe produces barely 1% of the soy consumed. Most of the world's production, just over 200 million tons, is carried out by three countries: the United States (43%), Brazil (34%) and Argentina (14%). However, some of this soy is produced from genetically modified crops, which can be a problem in terms of how the sector is perceived by socio-economic stakeholders and, more specifically, by consumers.

Soybeans are mainly used for their high protein content (36% of dry matter) and for the presence of an essential amino acid, lysine, which is not found in all plant resources such as cereals. Furthermore, their low cost currently makes them the most attractive source of vegetable protein on the world market.

In this very specific context of a virtual monopoly, what can algae offer as an input in the animal feed industry? From a biochemical and therefore nutritional point of view, there are species of algae whose protein content (Chapter 4, section 4.3) is close to that of soybeans and in some cases similar to that of soybean meal. This is particularly true of the red alga *Palmaria palmata*, whose protein content reaches 35% of dry matter when cultivated (Morgan et al. 1980). The species *Porphyra tenera* has a protein concentration that varies between 33 and 47% of the alga's dry mass. Its maximum content is close to the usual protein content of soybean meal (48% of dry matter). These two species also contain lysine levels ranging from 2.7 to 5.0% of the total amino acid fraction (Fleurence 2004).

Certain red algae therefore appear to be a potential source of protein and amino acids for animal feed. However, the availability of algae as a resource is extremely low compared to that of soybeans and the economic cost of producing these two species makes their use in mass livestock farming unfeasible. However, this observation only applies to the use of algae as a source of protein to supplement or even replace soybeans as the main source of vegetable protein in animal feed.

The analysis is more nuanced if we consider the use of algae as an original and varied source in the nutrition of animals, consumed by humans. It becomes particularly relevant when considering the farming methods used for marine animals (crustaceans, mollusks, fish) in the context of the development of aquaculture industries (Fleurence et al. 2012). "Why feed farmed fish with 'green feed' containing soy, rather than with 'blue feed' containing algae?" This question, which may seem provocative, has generated much debate, supported by research whose findings are interesting for the future of "algae as a future fish feed".

8.5.1. *Feeding aquaculture animals (fish farming, abalone farming, pen farming)*

The consumption of algae by herbivorous fish and the mechanisms of food selection with regard to algal biomass are relatively well known (Montgomery and Gerking 1980). Algae can be introduced into the diet of carnivorous fish if they are included in the feed formulation, generally in the form of meal incorporated into the final feed. The use of algal meal in the diet of farmed fish has been the subject of numerous studies (Mustafa and Nakagawa 1995) (Table 8.3).

Fish	Algae tested
Sebastes schlegelii (rockfish)	*Undaria pinnatifida*
Girella punctata (blackfish)	*Ulva conglobata*
Acanthopagrus schlegelii (blackheaded sea bream)	*Ulva pertusa*
Pagrus major (Red sea bream)	*Ulva pertusa*, *Undaria pinnatifida*, *Ascophyllum nodosum*, *Porphyra yezoensis*
Seriola quinqueradiata (amberjack)	*Laminaria digitata*
Sparus aurata (giltheaded sea bream)	*Ulva lactuca*, *Pterocladia capillacea*
Salmo salar (Atlantic salmon)	*Laminaria* sp., alginate premix
Oreochromis niloticus (Nile tilapia)	*Ulva* sp.

Table 8.3. *Examples of algae integration into the diet of certain fish species (Mustafa and Nakagawa 1995; Mustafa et al. 1995; Wassef et al. 2005; Ergün et al. 2009)*

Adding a small amount of algae meal to fish diets can have positive effects on growth, food assimilation, physiology, stress response, disease resistance and sometimes meat quality (Mustafa and Nakagawa 1995). In blackheaded sea bream, the incorporation of *Ulva* (green alga), *Ascophyllum* (brown alga) or *Porphyra* (red alga) flour at a level of 5% of the feed ration significantly increases the growth of the animals, the muscle volume and the efficiency of the food assimilation process (Table 8.4).

This last point is particularly important as improved food assimilation, particularly of nitrogen, contributes to a reduction in nitrogen emissions into the environment.

Furthermore, adding algae to the diet of fish has no negative effect on animal survival (Mustafa et al. 1995) and even promotes it in some cases. Indeed, studies have shown that adding algae (5% *Ulva* flour) to the diet of sea bream (*Pagrus major*) significantly improves the fish's defense mechanisms against the bacterium *Pasteurella piscicida*, which is a feared pathogen in fish farms (Satoh et al. 1987). This response is reflected in particular by an increase in the number of lymphocytes and by the activation of the complement pathway and certain associated activities such as spontaneous hemolytic activity.

Zootechnical criteria	Control diet (without algae)	*Porphyra* diet	*Ascophyllum* diet	*Ulva* diet
Increase in biomass (g)	1,595	2,281	1,957	1,816
Mass gain (g/fish)	10.9	13.3	12.1	11.2
Daily growth rate (%)	3.50	3.70	3.55	3.52
Food assimilation rate (%)	51.5	62.3	56.8	52.4
Protein deposition in muscle (%)	9.31	12.48	10.74	9.84
Survival rate (%)	77.8	87.8	84.0	84.3

Table 8.4. *Effect of an algae-based diet on zootechnical parameters of blackheaded sea bream farmed by aquaculture (Mustafa et al. 1995)*

The active compounds present in fish meal and stimulators of fish disease resistance are mainly poly- or oligosaccharides. Oral administration of fucoidan (50–200 mg.kg^{-1} body weight.day^{-1}) to young carp (*Cyprinus carpio)* for 4 weeks increases the fish's natural defense mechanisms (Kakuta 2004).

The incorporation of algal polysaccharides such as alginates in the form of premixes into salmon feed has also yielded extremely interesting results in terms of strengthening the animals' immune status. This was reflected in particular by a significant increase in the lysozyme content in the blood of fish fed a diet based on alginate premixes (Gabrielsen and Austreng 1998).

The response of animals to stress (biotic or abiotic) is also an important factor in fish production. The introduction of algae of the genus *Pterocladia* (red alga) (10%) or *Ulva* (green alga) (5%) into the diet of sea bream significantly improves the survival of animals subjected to abiotic stress (exposure to open air for 5 mins) (Table 8.5) (Wassef et al. 2005).

The impact of an algae-based diet on the quality of the final product, i.e. fish meat, is not always clearly established. It appears to vary depending on the fish species considered and the algae species used as nutritional inputs. It is generally neutral or may have a real influence on the sensory characteristics of the flesh. Algae are rich in various pigments, particularly carotenoids, which can produce a coloration of the skin and flesh that is attractive to consumers (Mustafa and Nakagawa 1995).

Type of diet	Survival rate (%)
Control diet	20
Ulva diet (5%)	40
Pterocladia diet (10%)	55

Table 8.5. *Effect of an algae-based diet on survival of sea bream subjected to abiotic stress (Wassef et al. 2005)*

Currently, the use of algae in fish feed can be justified by the favorable impact this feed has on the growth and health status of animals (e.g. improved immunity, increased stress resistance). Their preventive use in fish farming could limit or even eliminate the use of antibiotics, thus ensuring a better quality product for consumers.

The effects of algae-based feed on the efficiency of nitrogen assimilation by fish are also a major advantage for promoting this type of feed. By promoting the assimilation of protein nitrogen, this type of nutrition limits the production of urinary and fecal waste that pollutes pond water.

The use of algae as feed, as well as biological filters in aquaculture systems, is another aspect of the value of algae in a system for producing animals for human consumption. This concept is known as integrated aquaculture. There are many models for combining algae production as a purifier of fishpond water with the use of part of the algal biomass in the feed of aquaculture species.

There are examples of abalone farming coupled with sea bream farming, in which algae are used both as biological filters and as feed (Neori et al. 2000). In this system, effluent from abalone farming is concentrated to produce nitrogen-rich feed for the fish, and effluent from fish farming is purified by algae (*Ulva lactuca* and *Gracilaria conferta*). The algae biomass produced is then used as feed for abalone, which are grazing gastropods that feed on green or red algae.

Land plants, cereals and legumes do not have this inducing effect on the natural defenses of fish and will never play the role of biological purifiers in aquaculture production systems. In this context, it now seems appropriate to consider the economic feasibility of introducing marine plants into the production of marine animals for human consumption.

The use of algae in the feed of aquaculture animals has also been tested in mollusk and crustacean species. In the case of mollusks, the use of the red algae *Palmaria palmata* and *Grateloupia turuturu* in the diet of European abalone (*Haliotis tuberculata*) has been the subject of recent studies (Garcia-Bueno et al. 2016). The results obtained show that Dulse is a food of choice for abalone growth. The best gains in size and weight are observed with a diet composed of *P. palmata* or a mixture of the two algae in equal proportions (50/50) (Table 8.6). *Grateloupia turuturu* alone, as feed, appears to be less popular with the gastropod. However, this species produces antibacterial substances that are effective against the *Vibrio harveyi* strain, which is a formidable pathogen in abalone farms (Garcia-Bueno et al. 2014). In this context, the development of a feed based on *Palmaria palmata* incorporating an extract of *Grateloupia turuturu* seems to offer a real opportunity to promote the use of algae in the aquaculture production of this mollusk, which is highly prized by Asian and European consumers. Other studies based on diets containing *Gracilaria* and *Ulva* have demonstrated their effectiveness in promoting the growth of juvenile abalone (Fleurence et al. 2012).

Zootechnical parameters of abalone	*G. turuturu* diet	*P. palmata* diet	Mixed diet (50/50)
Initial size (mm)	42.9	42.0	42.9
Final size (mm)	47.3	52.7	54.2
Initial weight (g)	12.7	10.3	11.8
Final weight (g)	16.5	27.3	27.8

Table 8.6. *Effects of diets based on* Palmaria palmata *and* Grateloupia turuturu *on the growth of European abalone (adapted from Garcia-Bueno et al. 2016)*

Pen farming could also benefit from the use of algae in production systems, particularly in the context of preserving the health of livestock. A fucoidan extracted from the brown alga *Sargassum wightii* has been found to have antiviral activity against the white spot syndrome virus that affects shrimp in aquaculture. In the presence of this virus, animal mortality is 100% after nine days. It is only 40% after 21 days when fucoidan is administered at a concentration of 400 $mg.L^{-1}$ of nutrient solution (Sivagnanavelmurugan et al. 2012). The use of algae in pen farming is also of interest for the treatment of effluents generated by intensive aquaculture (Fleurence et al. 2012).

The incorporation of algae into the production of marine animals for human consumption is therefore possible. This integration can take place at several levels, namely in the feed of the animals produced, in the health preservation of the livestock raised or in the sanitation of the water in the production systems. In this way, algae could therefore indirectly enter the human diet.

8.5.2. *Feeding livestock*

A livestock animal or production animal is raised for its profitability. Sheep, cattle and pigs fall into this category. Poultry, animals with a much shorter life cycle, do not fall under this definition. Paradoxically, there are many examples of the use of algae in livestock feed.

Algae have been used for centuries as livestock feed. References to this use can be found in ancient Latin texts (*Bellum Africanum*, 45 BCE), which report that the Greeks, in times of fodder shortages, washed algae in fresh water and fed it to their livestock (Evans and Critchley 2014). Icelandic sagas also describe the use of algae as livestock feed (Delaney et al. 2016).

In Iceland, fresh algae are commonly used in feed for cattle, sheep and horses. Algae could also be dried and stored in barns for use as livestock feed during the winter. This latter use prevailed until the 1900s. Today, algae processing plants that produce flour for animal feed have taken over this traditional use. These plants rely mainly on geothermal energy to dry the algae biomass. Since the 1960s, an Icelandic plant has also been producing algae extracts for animal nutrition.

In Scandinavian countries (Norway, Finland and the Swedish island of Gotland), algae are used systematically or occasionally to feed ruminants, including calves and pigs (Delaney et al. 2016). The species *Alaria fistulosa, Laminaria bongardiana* and *Fucus evanescens* are used in the form of flour for more specific pig feed. In Finland, brown algae of the genus *Laminaria* or *Alaria* are generally used in livestock feed.

In France, and more particularly in Normandy, examples of livestock consumption of Dulse and Wrack, have been reported since the 1920s. The algae were included in the feed ration of stabled animals.

Apart from human intervention, algae can also be consumed spontaneously by farm animals. On the Norwegian coast, sheep have been feeding on algae for generations. This diet is thought to improve their digestive capacity compared to animals located further inland. In Scotland, a breed of small black sheep, similar to the famous Ouessant sheep, feeds exclusively on algae. These sheep, which live on the Scottish island of North Ronaldsay, can consume a wide variety of species (Table 8.7), but they show a dietary preference for *Laminaria digitata* and *Laminaria hyperborea* (Hansen et al. 2003). They can consume the whole fresh algae on the foreshore during the grazing period, or it can be provided by the farmer after washing.

Brown algae	
Alaria esculenta	Whole, washed ashore
Ascophyllum nodosum	Whole, fresh
Chorda filum	Whole, washed ashore
Fucus spiralis	Whole, fresh
Himanthalia elongata	Whole, washed ashore
Laminaria digitata	Washed or fresh fronds
Laminaria hyperborea	Washed or fresh fronds
Laminaria saccharina	Whole, washed or fresh
Pelvetia caniculata	Whole, fresh
Saccorhiza polyschides	Washed frond
Red algae	
Palmaria palmata	Whole, washed ashore
Green algae	
Codium fragile	Whole, fresh
Enteromorpha intestinalis	Whole, fresh
Ulva lactuca	Whole, fresh

Table 8.7. *Algae species consumed by sheep on North Ronaldsay Island (Hansen et al. 2003)*

These practices, linked to the insular or coastal nature of the herds, show that algae are completely suitable feed for the nutrition of farm animals. They also reveal that some animals, particularly those living on islands, had no other food source than algae biomass washed up or exposed at low tide.

Apart from livestock, algae have also been used in poultry feed, particularly for laying hens (personal communication, Joël Fleurence). In the late 1980s, a poultry farmer in Briochin (Côtes-d'Armor, France) used Ulva flour in chicken feed to improve the color of egg yolks. As Ulva are particularly rich in carotenoids, hens that consumed algae meal produced eggs with an attractive "yellow-orange" color for consumers. Recent studies (2014) have shown the benefits of using algae in the diet of laying hens, both for the effect of this diet on egg yolk characteristics and for maintaining animal health (Kulshreshtha et al. 2014). A diet based on *Chondrus crispus* or *Sarcodiotheca gaudichaudii* (red alga) (2% of the feed ration) not only increases the size and volume of the yolk, but also promotes the establishment of beneficial bacteria such as *Bifidobacterium longum* (4–14 times) and *Streptococcus salivarius* (4–15 times) in the intestinal microbiota. This red algae-based diet also reduces the presence of the pathogenic bacterium *Clostridium perfringens* in the hen's intestine.

More anecdotally, algae, due to their high mineral content, are also used in more elaborate feeds for horses, particularly racehorses (e.g. Equitonic®, NAF algae for horses, etc.).

However, the use of algae as a substitute for soybeans in livestock feed does not appear realistic. This is because the global annual soybean production is approximately 300 million tons, compared to 37 million tons of algae (wet weight). On the other hand, the integration of algae, even if only as a prebiotic feed ingredient, into the diet of certain livestock could be considered in the future by certain sectors wishing to enhance the quality of their animal production.

8.6. References

Daul, C.B., Morgan, J.E., Lehrer, S.B. (1993). Hypersensitivity reaction to crustacean and mollusks. *Clinical Review Allergy*, 11, 201–202.

Delaney, A., Frangoudes, K., Li, S.A. (2016). Society and seaweed: Understanding the past and present. In *Seaweed in Health and Disease Prevention*, Fleurence, J. and Levine, I. (eds). Academic Press, Elsevier, Amsterdam.

Ergün, S., Soyutürk, M., Güroy, B., Güroy, D., Merrifield, D. (2009). Influence of *Ulva* meal on growth, feed utilization and body composition of juvenile Nile tilapia (*Oreochromis nilocotus*). *Aquaculture International*, 17, 355–361.

Ermakova, S., Sokolova, R., Kim, S.M., Um, B.H., Isakov, V., Zvyagintseva, T. (2011). Fucoidans from brown seaweeds *Sargassum hornery*, *Eclonia cava*, *Costaria costata*: Structural characteristics and anticancer activity. *Applied Biochemistry and Biotechnology*, 164, 841–850.

Evans, F.D. and Critchley, A.T. (2014). Seaweed for animal production use. *Journal of Applied Phycology*, 26, 891–899.

Fleurence, J. (2004). Seaweed proteins. In *Proteins in Food Processing*, Yada, R.Y. (ed.). Woodhead Publishing Limited, Cambridge.

Fleurence, J. (2018). *Les algues alimentaires : bilan et perspectives*. Lavoisier Tec & Doc, Paris.

Fleurence, J. (2025). *The Use of Algae in Human Health: From Traditional Medicine to Molecules with Therapeutic Interest*. ISTE Ltd, London and John Wiley & Sons, New York.

Fleurence, J. and Ar Gall, E. (2016). Antiallergic properties. In *Seaweed in Health and Disease Prevention*, Fleurence, J. and Levine, I. (eds). Academic Press, Elsevier, Amsterdam.

Fleurence, J. and Guéant, J.L. (1999). Les algues : une nouvelle source de protéines. *Biofutur*, (99)80435-9. doi: 10.1016/S0294–3506.

Fleurence, J., Morançais, M., Dumay, J., Decottignies, P., Turpin, V., Garcia-Bueno, N., Jaouen, P. (2012). What are the prospects for using seaweed in human nutrition and for marine animals raised through aquaculture? *Trends in Food Sciences and Technology*, 27, 57–61.

Fleurence, J., Morançais, M., Dumay, J. (2018). Seaweed proteins. In *Proteins in Food Processing*, Yada, R.Y. (ed.). Woodhead Publishing, Sawston.

Gabrielsen, B.O. and Austreng, E. (1998). Growth, product quality and immune status of Atlantic salmon, *Salmo salar* L. fed wet feed with alginate. *Aquaculture Research*, 29, 397–401.

Garcia-Bueno, N., Decottignies, P., Turpin, V., Dumay, J., Paillard, C., Stiger-Pouvreau, V., Kervarec, N., Pouchus, Y.F., Marin-Atucha, A.A., Fleurence, J. (2014). Seasonal antibacterial activity of two red seaweeds, *Palmaria palmata* and *Grateloupia turuturu*, on European abalone pathogen *Vibrio Harveyi. Aquatic Living Ressources*, 27, 83–89.

Garcia-Bueno, N., Turpin, V., Cognie, B., Dumay, J., Morancais, M., Marin-Atucha, A., Fleurence, J., Decottignies, P. (2016). Can the European abalone *Haliotis tuberculata* survive on an invasive algae? A comparison of the nutritional value of the introduced *Grateloupia turuturu* and the native *Palmaria palmata* for the commercial European abalone industry. *Journal of Applied Phycology*, 28, 2427–2433.

Gutiérrez-Rodriguez, A.G., Juarez-Portilla, C., Olivares-Banuelos, T., Zepeda, R.C. (2017). Anticancer activity of seaweeds. *Drug Discovery Today*, 2, 434–447.

Hansen, H.R., Hector, B.L., Feldmann, J. (2003). A qualitative and quantitative evaluation of the seaweed diet of North Ronaldsay sheep. *Animal Feed Science and Technology*, 105, 21–28.

Holdt, S.L. and Kraan, S. (2011). Bioactive compounds in seaweed: Functional food applications and legislation. *Journal of Applied Phycology*, 23, 543–597.

Kakuta, I. (2004). Enhancement of the biodefence activity and improvement of physiological condition of fish by oral administration of algae fucoidan. *Suisanzoshoku*, 52, 413–420.

Kulshreshtha, G., Rathgeber, B., Stratton, G., Thomas, N., Evans, F., Critchley, A., Hafting, J., Prithiviraj, B. (2014). Feed supplementation with red seaweeds, *Chondrus crispus* and *Sarcodiotheca gaudichaudii*, affects performance, egg quality, and gut microbiota of layer hens. *Poultry Sciences*, 93, 2991–3001.

Marrion, O., Fleurence, J., Schwertz, A., Mamelouk, L., Ksouri, J., Villaume, C. (2005). Evaluation of protein in vitro digestibility of *Palmaria palmata* and *Gracilaria verrucosa*. *Journal of Applied Phycology*, 17, 99–102.

Montgomery, W.L. and Gerking, S. (1980). Marine macroalgae for fishes: An evaluation of potential food quality. *Environment Biology Fisheries*, 2, 143–153.

Morgan, K.C., Wright, J.L.C., Simpson, F.J. (1980). Review of chemical constituents of the red alga *Palmaria palmata* (Dulse). *Economic Botany*, 34, 27–50.

Mustafa, M.G. and Nakagawa, H. (1995). A review: Dietary benefits of algae as an additive in fish feed. *The Isreali Journal of Aquaculture*, 47, 155–162.

Mustafa, M.G., Wakamatsu, S., Takeda, T., Umino, T., Nakagawa, H. (1995). Effects of algae meal as feed additive on growth, feed efficiency, and body composition in Red Sea Bream. *Fisheries Sciences*, 61, 25–28.

Neori, A., Shpigel, M., Ben-Ezra, D. (2000). A sustainable integrated system for culture of fish, seaweed and abalone. *Aquaculture*, 186, 279–291.

Ngo, D.H. and Kim, S.K. (2013). Sulfated polysaccharides as bioactive agents from marine algae. *International Journal of Biological Macromolecules*, 62, 70–75.

Noguchi, T., Matsui, T., Miyazawa, K., Asakawa, M., Ijima, N., Shida, Y., Fuse, M., Hosaka, Y., Kirigaya, C., Watabe, K. et al. (1994). Poisoning by the red alga "Ogonori" (*Gracilaria verrucosa*) on the Nojima coast, Yokohama, Kanagawa Prefecture, Japan. *Toxicon*, 12, 1533–1538.

Satoh, K.I., Nakagawa, H., Kasahara, S. (1987). Effect of *Ulva* meal supplementation on disease resistance of Red Sea Bream. *Nippon Suisan Gakkaishi*, 57, 1115–1120.

Sivagnanavelmurugan, M., Marudhupandi, T., Palavesam, A., Immanuel, G. (2012). Antiviral effect of Fucoidan extracted from the brown seaweed *Sargassum wigthii*, on shrimp Penaeus monodom postlarvae against white spot syndrome virus. *Journal of the World Aquaculture Society*, 5, 697–706.

Synytsya, A., Kim, W.J., Kim, S.M., Pohl, R., Synytsya, A., Kvasnicka, F., Copikova, J., Park., Y.I. (2010). Structure and antitumor activity of fucoidan isolated from sporophyll of Korean brown seaweed *Undaria pinnatifida*. *Carbohydrate Polymers*, 81, 41–48.

Vo, T.S., Ngo, D.H., Kim, S.K. (2012). Marine algae as a potential pharmaceutical source for anti-allergic therapeutics. *Process Biochemistry*, 47, 386–394.

Wassef, E.A., El-Sayed, A.F.M., Kandeel, K.M., Sakr, E.M. (2005). Evaluation of *Pterocladia* (Rhodophyta) and *Ulva* (Chlorophyta) meals as additives to gilthead seabream *Sparus aurata* diets. *Egyptian Journal of Aquatic Research*, 31, 321–332.

Wells, M.L., Potin, P., Craigie, J.S., Raven, J.A., Merchant, S.S., Helliwell, K.E., Smith, A.G, Camire, M.E., Brawley, S.H. (2017). Algae as nutritional and functional food sources: Revisiting our understanding. *Journal of Applied Phycology*, 29, 949–982.

Conclusion

Algae have been part of human and animal diets for centuries. Consumption by coastal populations dates back to prehistoric times. Its regulatory status as a "non-traditional food" is therefore not justified in light of historical and ethnographic data. The difference in the perception of algae as food is, in fact, closely linked to the geography and history of human populations.

Coastal and, even more so, island populations (e.g. Japan) have been able to take advantage of the marine resources, both animal and plant, available to them, as terrestrial agricultural resources were often limited. Animals living in similar conditions, such as the famous sheep of the North Ronaldsay Islands, have followed the same dietary adaptation behaviors.

In Europe, algae were seen by coastal populations as a food of famine, and this negative image has long prevailed in the collective unconscious. This image is all the more damaging given that there are many edible species of algae and that this resource represents an irreplaceable nutritional source of unique soluble dietary fiber and trace elements of major nutritional interest (e.g. Fe, Mg, Ca).

The presence of all eight essential amino acids in algae is also an advantage for consumers who have chosen to adopt diets that exclude animal proteins (e.g. vegans). The presence in algae of molecules (e.g. polyphenols and polysaccharides) with anti-inflammatory and even anti-allergic properties also opens up the possibility of developing functional food products free of major allergens. Although this market niche appears limited, it could meet the demand of consumers suffering from serious food allergies.

The intensive development of water-consuming crops (e.g. soybeans, rapeseed and maize) also raises questions about how humanity will manage freshwater reserves in the not-too-distant future, i.e. within a century. In this context, greater use of marine plant resources such as algae, which do not require freshwater for their production, appears to be a promising avenue for future exploration. Furthermore, algae are an easily renewable biological resource, provided that production conditions (cultivation, harvesting) respect the biological and ecological cycles of these marine plants. They are therefore easily renewable foodstuffs that are useful in animal and human nutrition.

In this age of globalization and cultural and culinary diversity, we might question the fact that all animals intended for human consumption, whether terrestrial or aquatic, are fed almost exclusively on a single plant resource. The diversity of food available to us is therefore only apparent. Given this situation, greater use of algae in animal feed and, indirectly, in human food is an option that deserves socio-economic consideration and perhaps new societal choices.

Index

E, F

G, H

I, K

L, M

O, P

R, S

T, U

V, W

X, Y, Z

Printed and bound by CPI Group (UK) Ltd, Croydon, CR0 4YY

13/09/2026

14961010-0001